T0140721

The Deutsche Nationalbibliothek lists this publication in the Deutsche Nationalbibliografie; detailed bibliographic data are available in the Internet at http://dnb.d-nb.de .

ISBN 978-3-8325-3527-8

Logos Verlag Berlin GmbH
Comeniushof, Gubener Str. 47,
10243 Berlin
Tel.: +49 (0)30 42 85 10 90
Fax: +49 (0)30 42 85 10 92
INTERNET: http://www.logos-verlag.de

Novel Haptic Cues for UAV Tele-Operation

Dissertation

Ph.D Course on:
Automation, Robotics and Bioengineering

University of Pisa
Department of Energy and Systems Engineering
and
Max Planck Campus Tübingen
Max Planck Institute for Biological Cybernetics

Samantha M.C. Alaimo

2013

Typeset by the author with the $\LaTeX$ Documentation System.

Author email: samanthaalaimo@yahoo.it

Date of examination:	July 15, 2011
PhD Course Chairman:	Prof. Dr. A. Caiti
MPI Director:	Prof. Dr. H. H. Bülthoff
Tutor:	Prof. Dr. L. Pollini
Examination board:	Prof. Dr. M. Innocenti
	Prof. Dr. P. Castaldi
	Prof. Dr. A. Giannitrapani

Ackowledgements

This thesis summarizes the research mainly conducted at the Max Planck Institute (MPI) for Biological Cybernetics in Tübingen which is under the direction of Professor Heinrich H. Bülthoff. Special thanks to Professor Heinrich H. Bülthoff and to my supervisor, Professor Lorenzo Pollini. Professor Heinrich H. Bülthoff not only gave me the financial support and provided me with the MPI facilities but also imparted really important advices on my work in which he was absolutely interested in and his constant presence proves it. My advisor, Professor Lorenzo Pollini, and the University of Pisa gave me the sponsorship and the chance of starting such an important work experience in Tübingen. Although Lorenzo was in Pisa, his Skype-presence and his fantastic problem solving characteristic were significant for my work.

Thanks to Jean Pierre Bresciani for making available his excellent experience with the experiment design and the results' analyses at each hour of the day.

I also acknowledge Professor Alfredo Magazzù for being the most important reference point on the aircraft aerodynamics, regularly ready to answer to my questions and happy to aid with his always useful and outstanding instructions.

Thanks to Marco Locatelli, Giuseppe Pasanisi and Alberto Cioffi,

professional and test pilots, for their useful suggestions and their availability every time I needed.
Thanks to Pierluigi Capone, Head of Control Laws Design in AgustaWestland, for recommending important references.
Thanks to Professor Alonge for having strongly wanted me to make the PhD studies and for providing the knowledge needed and to Professor Grillo for suggesting useful advices and references although the distance.
Thanks to all the participants to the experiments professional and naïve pilots who, enjoying the “video-game”, gave me the chance to make the experiments even in the evening and during the weekend.
Thanks to MPI and University of Pisa for giving the chance to attend important classes and PhD summer schools (in Siena, Pisa, Bertinoro, Munich) in which I met big names in non-linear systems, simulators and teleoperation fields: thanks to Prof. Alberto Isidori, Prof. Alessandro Astolfi, Prof. Lorenzo Marconi, Prof. Frank Cardullo, Prof. Günter Niemeyer, Prof. Keyvan Hashtrudi-Zaad, Prof. Yasuyoshi Yokokohji, Prof. Edward Colgate, Prof. Manuel Ferré and Prof. Angelika Peer for providing a great amount of knowledge.
The MPI in Tübingen does not only offer an amazing amount of selected devices and some of the best researchers ever but also a really nice and friendly environment. I will never forget the wonderful time spent with my MPI colleagues and, above all, friends.
Thanks to my best friends and colleagues who supported me a lot till the deadline of the thesis submission.
Last but not least thanks to my whole family for giving me all the support I needed. Without their understanding and love this thesis would not have been possible.
Although this period was very hard and tiring, I actually enjoyed my years as a researcher and I am happy having done it.

Contents

List of Figures

List of Tables

Abstract

The use of Unmanned Aerial Vehicles (UAVs) is continuously increasing for both military and civilian operations. The degree of automation inside an UAV has reached the capability of high levels of autonomy but human participation/action is still a requirement to ensure an ultimate level of safety for the mission. Direct remote piloting is often required for a board range of situations; this is particularly true for larger UAVs, where a fault might be dangerous for the platform and even for the other entities of its environment (people, buildings etc.). Unfortunately the physical separation between pilot/operator and the UAV reduces greatly the situational awareness; this has a negative impact on system performance especially in the presence of remote and unforeseen environmental constraints and disturbances. For this reason the present dissertation addresses the study of means to increase the level of situational awareness of the UAV operator.

The sense of telepresence is very important in teleoperation, it appears reasonable, and it has already been shown in the scientific literature that extending the visual feedback with force feedback is able to complement the visual information (when missing or limited). An artificially recreated sense of touch (haptics) may allow the operator to better perceive information from the remote aircraft state, the environment and its constraints, hopefully preventing dangerous situations. The dissertation introduces first of all a novel classification for haptic aid systems in two large classes: Direct Haptic Aid (DHA) and Indirect Haptic Aid (IHA), then, after showing that almost all existing aid concepts belong to the first class, focuses on IHA and tries to show that classical applications (that use a DHA approach) can be revised in a IHA fashion. The novel IHA systems produce different sensations, which in most cases may appear as exactly "opposite in sign" from the corresponding DHA; such sensations can provide valuable cues for the pilot, both in terms of performance improvement and "level of appreciation". Furthermore, the present dissertation shows that the novel IHA cueing algorithms, which were designed just to appear "natural" to the operator and not to directly help the pilot in the task (as in the DHA cases), can outperform the corresponding DHA systems.

Three case studies are selected: obstacle avoidance, wind gust rejection, and a combination of the two. For all the cases, DHA and IHA systems are designed and compared against baseline performance with no haptic aid. Test results show that a net improvement in terms of performance is provided by employing the IHA cues instead of both the DHA cues or the visual cues only. Both professional pilots and naïve subjects were asked to test them through a deep campaign of experiments. The perceived feelings transmitted by the haptic cues, strongly depend by the type of the experiment and the quality of the participants: the professional pilots, for instance, retained the DHA the most helpful force while they preferred IHA because they found it more natural and because they felt a better control authority on the aircraft; different results were obtained with naïve participants who in the obstacle avoidance task retained both the haptic aids the most helpful forces in harmony with the results being although divided on their own preferences, while in the windy obstacle avoidance task they retained the visual cue the most helpful (despite the results) one and preferred it as well because maybe not enough trained on the haptic aids.

In the end, this thesis aim is to show that the IHA philosophy is a valid and promising alternative to the other commonly used, and published in the scientific literature, approaches which fall in the DHA category.

Finally the haptic cue for the obstacle avoidance task was tested in the presence of time delay in the communication link as in a classical bilateral teleoperation scheme. The Master was provided with an admittance controller and an observer of force exerted by the human on the stick was developed. Experiments have shown that the proposed system is capable of standing substantial communication delays.

1 Introduction

1.1 Unmanned Aerial Vehicles

Unmanned Aerial Vehicle (UAV) is the name commonly used to describe an airborne vehicle without any pilot on-board, which operates under either remote or autonomous control. UAVs are also referred as Remotely Piloted Vehicles (RPVs), Remotely Operated Aircrafts (ROAs), Unmanned Vehicle Systems (UVSs) or simply Drones. In most instances, the term RPV might be more appropriate suggesting that the vehicle is remotely controlled and still relies, to a great degree, on human involvement.

UAVs are mainly employed in military field. Lessons from recent combat experiences in Kosovo, Afghanistan and Iraq have shown that UAVs can provide vastly improved acquisition and more rapid dissemination of Intelligence, Surveillance and Reconnaissance (ISR) data [55]. Over the past several years, a confluence of events and developments has brought the Military Services to change the way of perceiving the UAVs. These include:

- Dramatic increase in computer processing power;
- Advances in sensor technologies that reduce sensor size and weight, provide high resolution, and permit detection of fixed and moving targets under a variety of environmental conditions;
- Improved communications, image processing, and image exploitation capabilities.

UAVs have the potential to reduce operational and support cost as compared to the use of manned aircraft [14].

Currently UAVs have a permanent position in the military arsenal in the US, Europe, Middle East and Asia. Today UAV development strives toward more peaceful and civil usage [53] such as rescue, border surveillance, disaster monitoring, telecommunications relay, fire fighting, traffic monitoring, pipeline surveillance, agriculture, construction, and public utility operations [35]. Thus, police, forest rangers, fire brigades are very interested on them for public security. UAVs civil employment also includes video-taping for photogrammetric or scientific applications [13].

Communications represent the most important subsystem for UAVs. Bandwidth is needed to support systems that control the UAVs flight, launch and recovery, to transmit the output of on board sensors to both line of sight and beyond line of sight processing centers, and to communicate with air traffic control centers. Equally important is the recognition of a mission area for UAVs acting as communication relays linking tactical forces, including other UAVs, and providing connection to support centers.

The potential benefits of UAVs, such as low operational cost and no risk of losing human lives, make sense when the teleoperation is safe and no mishaps and accidents occur. A crash of a UAV during teleoperation could not only lead to possible damage to the local environment but also to the loss of the vehicle. Humans in the vicinity of the incident may get injured as well. Therefore, safety in UAV teleoperation is of great importance not only for mission success but also to preserve the sustainability of UAV operations [39].

1.2 Manual vs autonomous control

Various ways to control UAVs exist. They can be categorized in autonomous control and manual control.

Some of the problems associated with the automatic control are [53]:

- Reduced situation awareness;
- Increased monitoring demands;
- Cognitive overload;
- Mis-calibration of trust in automation (either excessive trust, termed "complacency", or, at the other extreme, mistrust of automation);
- Inability to re-assume manual control;
- Degraded manual skills through lack of practice;
- The need for new selection and training procedures;
- Increased inter-operator coordination requirements;
- Increased workload management requirements;
- Loss of motivation and job satisfaction;
- Increase in the risk of human error because of the human weakness to maintain vigilance during extended periods of relatively low task demand.

Furthermore, fully-autonomous systems are more suitable for simple

missions with, for example, pre-defined targets and far away from inhabited environments.
Manual teleoperation could enable instead more flexibility in controlling a UAV close to inhabited environments and without predefined targets [39]. Focusing on manual control would give to the pilot the freedom to choose the targets step by step (for example because of last minute communication from control towers). Furthermore, the complex scenarios in which UAVs may operate require the presence of the human operator in the decision making process.

For all these reasons, keeping a human operator in-the-loop is more advisable.

1.3 UAV Mishaps

There are several factors at work contributing to UAV mishaps.

Besides electro-mechanical failures (62%), mishaps and incidents in UAV teleoperation are, for a great part, due to human errors during operation (25%) [14]. This is essentially due to the lack of the natural, multiple-sensory information of the environment. In fact, the remote pilot, inside the Ground Control Station (GCS, see Figure 1.1), is characterized by the following troubles:

- Limited Field Of View (FOV) cameras (i.e. no "look around" possibility, etc.);
- No inertial cues (motion, vibrations, gravity/attitude etc.);
- No auditory cues;
- Video/data communication delays;
- No feedback on control stick of the environment around the remote

Figure 1.1. UAV remote piloting from a Ground Control Station (picture from http://www.flickr.com).

vehicle (obstacles, disturbances etc.).

Usually, in order to solve the first mentioned trouble, the UAV operator is supplied with a richer visual information like showing different cameras on various displays. Another alternative is to supply the operator with a continuously updated "augmented reality" or "synthetic vision" produced by a computer resembling reality [47]. As concerning the inertial cues, some steps on the employment of motion cues to augment UAV operator performance and improve UAV flight training was made [29, 60]. About the auditory cues, augmented reality through multi modal tactile and auditory information displays are used in other fields to resemble reality [53, 54]. The communication delays, depending on the situation, turn out in the range of 100 to 1600 ms (and even more). This is a considerable amount given that 100 ms delay usually leads to measurable degradation of human performance [8, 74]. Delays of about 250-300 ms quite often lead to unacceptable airplane handling qualities [66]. Some techniques were used in the past to enhance the performance of a teleoperator in presence of time delay; for instance, automatic switching for stopping override [59] or the use of the predicted display [62]. As concerning the haptic feedback, tactile cues have shown to complement the visual information (through the visual displays of a remote GCS) and upgrade the efficiency of the UAV teleoperation [47, 40, 53].

In conclusion, augmented feedback to the operator such as haptic feedback and multi modal displays can compensate, to some extent, for the lack of sensory cues that would be presented to UAV operators [53]. Introducing the mentioned augmented feedbacks in the GCS would hopefully imply a reduction of the UAV mishaps.

Thus, investing in a human machine interface design tailored on the human needs would upgrade the operator situational awareness (see later) and maybe the performances.

1.4 Situational Awareness

By the late 1980s, there was a growing interest in understanding how pilots maintain awareness about the many complex and dynamic events that can occur simultaneously in flight, and how this information was employed to guide future actions. The vast quantities of sensor information available in the modern cockpit, coupled with the flight crew's "new" role as a monitor of aircraft automation, increased interest on Situational Awareness (SA) issue [61]. Through the word "situation(al) awareness", the processes of attention, perception, and decision making that together form a pilot's mental model of the current situation of the aircraft is described [18]. According to [17], the crew's knowledge of both the internal and external states of the aircraft, as well as the environment in which it is operating is defined as SA.

In fact, the 'health' of its utility systems (the internal state of the aircraft) and terrain, threats, and **weather** (the external state) must be monitored.

To expand upon this definition, Endsley [19], described the three hierarchical phases of SA: perception, comprehension, and projection. The First SA Level, named *Perception of the elements in the environment*, includes perceiving the status, attributes, and dynamics of relevant elements in the environment (airspeed, position, altitude, route, direction of flight etc) and also weather, air traffic control clearances, emergency information etc. [19]. The Second SA Level, named *Comprehension*, is based on an understanding of the significance of the First SA Level elements. The Third SA Level, named *Projection*, is based on the knowledge of the status and dynamics of the elements and a comprehension of the situation (both First and Second SA Levels).

SA is not synonymous with good performance. In fact, having good SA might bring good performance: a pilot could have a good SA without being a good pilot for the lack of motor skills, because of co-ordination or attitude problems etc. Conversely, under automatic flight conditions it is possible to have good performance with minimal SA [64].

As concerning SA in automation, SA is something that a person creates himself through perception (First SA Level) and it could not be provided by automation which usually excludes the human operator from the control loop, though automation can be thought in a different way say *supporting SA through decision aids and system interfaces*. And **SA can be hindered if automation designers fail in adequately addressing the SA need to the operator** [64].

Since SA is created through the perception of the situation (Level 1), the quality of SA is very dependent on how the person directs attention and how attention to information is prioritized based on its perceived importance. Jones and Endsley (1996) [36] found that operators were prone to overlooking crucial information in sustaining SA, though all relevant and needed information was present. Actually, this was found to be the most frequent causal factor associated with SA errors [53].

The above definitions are written in case of aviation in general but can be extended to the case of UAV teleoperation as long as the GCS is, in this case, fixed to the ground. Thus, as seen in Section 1.3, being aware of the aircraft internal and external state is much more difficult for the pilot. According to [53] haptic feedback can compensate to some extent for the lack of sensory cues that will be presented to UAV operators (see Section 1.3), this means that the addition of a haptic interface to the visual interface may improve the situational awareness of a remote UAV pilot and the efficiency of the teleoperation.

1.5 Bilateral Teleoperation

In teleoperation, a human operator conducts a task in a remote environment via master and slave manipulators [8].

One of the advantages of a teleoperation system is to combine the human capabilities with the robot ones. UAVs have also been referred to as non-anthropomorphic robots [76]. Through the teleoperated systems barriers like distance, hazardness or scaling can be overcome.

Remote teleoperation can be classified into unilateral and bilateral. In unilateral teleoperation no haptic feedback is available to the operator. In bilateral teleoperation, haptic feedback allows the operator to have a better feeling about the remote environment, providing a more extensive sense of telepresence [24].

The word *telepresence* refers to an experience that appears to involve displacement of the user's self-perception into a computer-mediated environment [16]. In particular the word telepresence is employed when the remote environment is real and not synthetic. In this case it is instead referred as *virtual presence* [16].

In particular, in a haptic/bilateral teleoperation system, a human operator controls a remotely located robot or slave device via a human-machine interface or master device while receiving haptic feedback of the interaction between the teleoperator and the (virtual or real) environment.

Stimulating a human's sense of touch by managing with sensation of movement or force in muscles, tendons, and joints is referred to as having a kinesthetic or haptic sensory experience [70].

As haptic data from the master site enters the control on slave site and vice versa, a control loop between the subsystems human-master and slave-environment is closed over the communication channel. This poses several challenges for control design, above all in the presence of time delay in the communication links (see Section 2.4).

1.6 Goal of the Thesis

The aim of this work is the investigation of possible haptic aids for teleoperated systems. In particular this thesis focuses on the teleoperation of UAVs. The principal issue of remote piloting an UAV is represented by the physical separation between pilot and vehicle which causes an almost complete absence of the sensorial information usually available when on board.

The purpose of this report is threefold. First, it presents a novel classification of the haptic aids present in literature in two classes: Indirect Haptic Aids (IHA) and Direct Haptic Aids (DHA) (see Chapter 2). This is a contribution on the research on the enhancing the UAV pilot Situational Awareness. In fact, by assuming that haptic aids provide an improvement of the SA, this thesis launches a highly important challenge that is to explore which haptic feedback philosophy should be followed in order

to better improve the SA. In particular, the main goal of this thesis is to show that the Indirect Haptic Aid philosophy is a valid alternative to the other commonly used, and published in the scientific literature, approaches which mainly fall in the Direct Haptic Aid category. Second, it investigates the potential of using a novel concept of tactile interaction as an information source of the external conditions of the air bone aircraft. Third, it explores the benefits of multi-modal information sources on the flight deck, in terms of improving attention and enhancing flight performance in presence of communication delays as well.
This work focuses on the investigation of possible haptic cues meant to improve the virtual immersion of the remote pilot. Three novel haptic feedbacks were designed. The first one is a reality-inspired haptic aid since it was built to transmit to the UAV operator a realistic situation which is happening outside the aircraft: the external disturbances such as wind gusts. The second one is an artificial component since it depends on environmental constraints. The third one is both a reality and a virtual reality-inspired haptic aid and it merges the first two haptic feedbacks.

The haptic feedbacks will be provided to the human operator via a haptic control device. As concerning the reality-based haptic feedback, the research resulted in the *Conventional Aircraft Artificial Feel*. As concerning the artificial-based haptic feedback, the research resulted in a novel philosophy of an obstacle avoidance haptic feedback, the *Obstacle Avoidance Feel*, which was built to help the UAV operator in detecting and hopefully avoiding the obstacles. As concerning the mixed reality/virtual reality-based haptic feedback, the research resulted in the *Mixed Conventional Aircraft Artificial Feel/Obstacle Avoidance Feel* which extends the previously described haptic aid systems by merging them into a system capable of aiding a pilot involved in a flight within a constrain environment in the presence of wind gusts.

All of the just introduced haptic feedbacks fall in the class of Indirect

Haptic Aids. The mentioned Conventional Aircraft Artificial Feel will be shown to increase the performance in terms of instinctive response to a stimulus in pilots without any previous training on the experiment. It also improves the situational awareness intended as making the pilot to feel as piloting the aircraft on board. The Obstacle Avoidance Feel and the Mixed Conventional Aircraft Artificial Feel/Obstacle Avoidance Feel will be shown to provide a net improvement in the operator sensation with respect to the existing obstacle avoidance haptic aids belonging to the Direct Haptic Aids class. This would improve the safety of the teleoperation by keeping higher the attention of the pilot in the task and thus enhancing the situational awareness.

1.7 Thesis outline

The structure of this report is the following: Chapter 2 presents a review about the haptic aids published in scientific literature and classifies them in two categories: Direct Haptic Aid (DHA) and Indirect Haptic Aid (IHA). It also shows the problem of the presence of delay in the communication link of a bilateral teleoperation and it mentions the remedies proposed in literature. Chapter 3 describes in details the Conventional Aircraft Artificial Feel (CAAF) which, as will be shown, belongs to the IHA class. The newly introduced CAAF haptic force was evaluated and Section 3.6 shows the evaluation results. Chapter 4 describes in details the Obstacle Avoidance Feel (OAF) which, as will be shown, also belongs to the IHA class. The newly introduced OAF haptic force was evaluated and Section 4.5 shows the evaluation results. Chapter 5 presents and evaluates the Mixed Conventional Aircraft Artificial Feel/Obstacle Avoidance Feel (Mixed-CAAF/OAF), belonging to the IHA-class as well. It was evaluated as well and Section 5.6 shows the experimental results. Finally, the Chapter 6 considers the introduction of the time delay in the communication link and proposes the application of an admittance-control scheme for the master side with the new introduction of an observer to estimate the human operator force in case of lack of force sensors in the employed haptic device.

2 Haptic Systems Review and Classification

As mentioned in Chapter 1, in a general teleoperation setting, the human exerts a force on the master manipulator; it results in a displacement which is transmitted to the slave that mimics the movement of the master. If the slave possesses force sensors, then it can transmit, or reflect back to the master, the reaction forces from the task being performed in the remote environment; these enter into the input torque of the master, and the teleoperator (as comprised of two robotic subsystems a master and a slave that exchange signals as positions, velocities and/or forces) is said to be controlled bilaterally (see Figure 2.1) [31].

Although reflecting the encountered forces back to the human operator enables the human to rely on his/her tactile senses along with visual senses, it may cause instability in the system if delays are present in the communication media. This delay-induced instability of force reflecting teleoperators has been one of the main challenges faced by researchers [74, 8, 52, 63, 11, 66].

The teleoperation through haptics has already 50 years of history. Indeed, in 1950 the first master-slave teleoperator was built by Goertz [26]

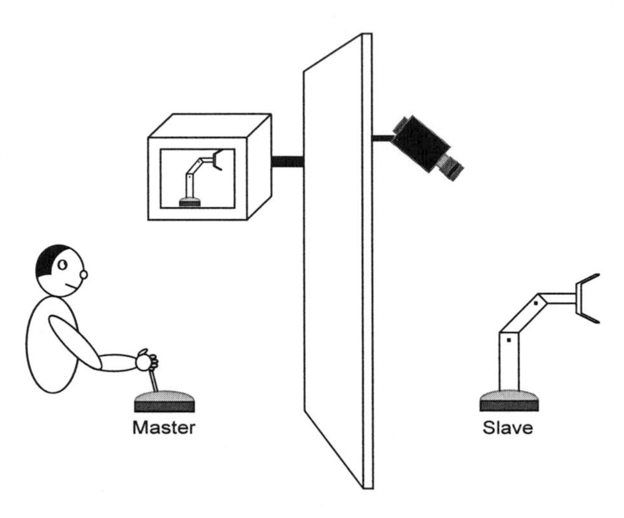

Figure 2.1. Bilateral teleoperation.

to remotely handle radioactive substances. Since that work, the number and diversity of teleoperation applications have considerably increased. Today, such systems are used in underwater exploration, manufacturing, chemical and biological industry, and, more recently, in the medical field. This Chapter focuses in the most recent application: the mobile robot teleoperation.

According to [53] haptic feedback can compensate to some extent for the lack of sensory cues that are presented to UAV operators (see Subsection 1.3), this means that the addition of a haptic interface to the visual interface may improve the situational awareness of a remote UAV pilot and the efficiency of the teleoperation. It is particularly necessary in case of limited visual informations. In the presence of foggy weather conditions, for example, or because of the employment of a limited FOV camera, the haptic feedback provides information through the sense of touch, which can be applied directly on the control device. It is well known that the reaction to the perceived haptic information is faster (3 Hz) with respect to visual information (0.5 Hz). This is due to the spinal cord that acts as a subconscious fast controller [65].

In the next subsection a review of the mobile robot teleoperated systems is presented.

2.1 Robot Bilateral (Tele)operation Review

Some of the numerous applications of teleoperation are operating space robots from ground, commanding unmanned underwater vehicles, handling hazardous materials, maneuvering mobile robots with obstacle avoidance. The present section focuses on the teleoperation of mobile robots.

The following subsections review the Ground Mobile Robots and Manned/Unmanned Aerial Vehicles bilateral teleoperations.

2.1.1 Ground Mobile Robots

This subsection presents a review about the teleoperation of ground mobile robots. Reference [15] makes use of a haptic interface in order to increase the user's perception of the workspace of the mobile robot. In particular, a virtual interaction force is computed on the basis of obstacles surrounding the mobile vehicle in order to prevent dangerous contacts, so that navigation tasks can be carried out with generally better performances. When an obstacle is close enough to the mobile robot it exerts a spring damper virtual force on the human operator through the haptic device in order to help him/her in avoiding the collision with the obstacle.

Also in [20] the force feedback is based on measured distances from the mobile robot to the obstacles. The force feedback gain is variable based on measured distances to the obstacle and derivatives of the distances. Clearly, the gain is higher when the obstacle and the mobile robot approach each other than when obstacle and robot are moving away from each other.

In [32] the goal location exerts an attractive force on the teleoperator which is proportional to the distance between the goal location and the mobile robot.

References [15, 20, 71] make use of the *Car-Driving Metaphor* which utilizes position-velocity kinematic mapping: the displacement of the end-effector of the haptic device is mapped to the linear and angular velocities of the mobile robot. A 3D approach of the car-driving metaphor is presented in [33]: the Intuitive Haptic Conical Control Surface. Here, the third vertical coordinate provides the current velocity of the robot and so the conical surface allows intuitive haptic detection of the zero speed. For example, a force directed to the zero speed point (the cone's vertex) is a suggestion to the human operator to decrease the commanded velocity of the mobile robot.

Also in [71] the obstacle force feedback exerted on the human operator is a repulsive one and it is proportional to the distance between the robot and the obstacles.

2.1.2 Manned and Unmanned Aerial Vehicles

The present section presents a literature review concerning operation (remote or not) of aerial vehicles, both manned and unmanned. In [84], 68 actuators form a vibrotactile image that can be updated in real-time navigation, hovering, threat warning, spatial disorientation countermeasures, communication, etc. The actuators are attached to the body and communicate information by vibrating at a specific location. The most simple set-up is when only one actuator vibrates: it is attached to that side of the body that corresponds to the desired direction of movement. Possible applications in land (navigation support and threat warnings for drivers, infantrymen, blind people, etc.), underwater (divers), and in space (astronauts in the International Space Station).

Reference [40] investigates the application of haptic feedback in UAV teleoperation for collision avoidance in low airspace by mapping of the environmental constraints that can even be outside the visual FOV. In the context of teleoperated systems where visual cues only have usually been used, the adoption of an artificial feel system for the stick appears to increase the situational awareness; this is extremely relevant for UAVs.

Tactile cues have shown to complement the visual information (through the visual displays of a remote GCS) and improve the efficiency of the teleoperation [40]. The task of the experiment is to fly from waypoint to waypoint as accurately as possible in an obstacle-laden environment. Stick deflection tilts the swashplate (as in a real helicopter). The force on stick is proportional to the distance between the UAV and the obstacles. Through a rather complex remote piloting and obstacle avoidance simulations, [40] shows that an appropriate haptic augmentation may provide the pilot with a beneficial effect in terms of performance in its task: the authors, in fact, extensively studied the problem of force feedback (injecting an artificial force on the stick) and stiffness feedback (changing stick stiffness to oppose less or more strongly to motion). The active deflection of the stick given from the force feedback can be considered an "autonomous collision avoidance" function. In fact, the force feedback can be regarded to yield a "commanded" stick deflection that the operator should follow as much as possible. That is, when yielding to the forces applied on the hand, the operator deflects the stick in a way that satisfies the collision avoidance function. With stiffness feedback instead, the stick becomes stiffer when in the presence of an obstacle, that is: the extra stiffness provides an impedance resulting in an extra force which depends on the deflection of the stick by the operator. The authors of [40] then conclude that a mixed force-stiffness feedback is the best solution. This type of haptic augmentation systems for RPVs were designed in order to help directly the pilot in his/her task by pulling the stick in the correct direction for the achievement of the task, or by changing stick stiffness in order to facilitate or oppose to certain pilot's actions [43, 40].

Another work not about teleoperation but still about haptic augmentation is the one by De Stigter [79]: he suggests to use the haptic device similarly to the artificial horizon with flight director: as bringing the artificial horizon bar in the center would let the aircraft to fly in the desired direction, by bringing the haptic device to the central position the target path will be followed in a close future. In fact, the haptic device moves in the opposite direction with respect to the one required by the target path and about a quantity proportional to the future error between the actual path and the one to follow.

Reference [50] proposes the introduction of an active stick in a manned military aircraft (Alenia Aermacchi M-346). In training aircrafts, the introduction of an active stick in each cockpit would be very useful as long as the two sticks can be electrically connected; thus they could work in a synchronous way as they were mechanically connected. In this way, the trainer gets the chance to supervise the control input of the apprentice pilot. The trainer could also make little corrections to teach the best way to impart some maneuver to the aircraft. The active stick would move also coherently with the autopilot commands to inform the pilot about the approaching of the envelope limits (already present in fly-by-wire aircrafts through the *stick shaker*). This is in line with what is stated in [51]: the active stick in this case makes the system structure and the automation processes visible to the operator. This aid in identifying options for action can help the operator in maintaining SA.

2.2 Haptic aids analysis and classification

Most of the described papers focus on a collision avoidance support to help the pilot in avoiding obstacles. Usually this kind of haptic aids, for example, have always been represented by repulsive forces created by objects in the environment in order to help the operator in escaping the collision with them.

When the task is instead a path to follow, a target location to reach or a desired stick position to get, the haptic feedback is instead attractive with respect to the task.

Thus, in all the described papers except for the [79], the haptic force that is artificially injected in the stick has the same sign (i.e. direction) as the one needed in order to achieve the requested task; thus the operator has to be compliant with it in order to avoid the obstacles or to reach the desired position.

As concerning the work [79] instead, the haptic force has the opposite sign with respect to the one desired in order to achieve the requested task and the human operator has to appose the force exerted from the stick by keeping the stick in the center while the haptic force tries to move it away on the sides.

Due to the last considerations, the haptic force used in the bilateral teleoperation of RPV can be divided (as the author published [1, 2, 3, 4, 5, 6, 7]) in two philosophies: Indirect Haptic Aiding (IHA) versus Direct Haptic Aiding (DHA):

- **Direct Haptic Aid**: the class of all Haptic aids which produces forces and/or sensations (due to stick stiffness changes for instance) aimed at "forcing" or "facilitating" the pilot to take some actions instead of others. The operator has to be compliant with the force felt on the stick to achieve the task;

- **Indirect Haptic Aid**: the class of haptic aids where the sense of touch is used to provide the pilot with an additional source of information that would help him/her, indirectly, by letting him/her know what is happening in the remote environment and leaving him/her the full authority to take control decisions. In general, in this case the operator has to oppose to the force felt on the haptic device.

It is clear from the above definitions that these two classes of haptic aids are complementary.

In practice under DHA, the haptic feedback suggests the correct direction the pilot should move the stick in order to achieve the task and the operator has to be compliant with it, while under IHA the haptic feedback is, in general, in the opposite direction and the operator has, in general, to oppose to it.

The stretch reflex, which is a reflex contraction of a muscle in response to passive longitudinal stretching, is an highly automatic motor response that is believed to be the spinal reflex with the shortest latency [73]. The author believes that the stretch reflex is involved when using IHA-based haptic feedback. Thus, a strength point of IHA is that, as a matter of fact, when a haptic input requires a reaction to a stimuli rather than compliance, it might be more "natural" for the human being [73, 38].

Another difference between the two classes is the behavior of the system with the pilot out of the loop: the DHA approach closes the loop itself being it an "almost-automatic-system-concept". The IHA approach instead, as will be clarified later, is more likely to produce a system that requires the presence of an operator in the loop in order to achieve the task. As a matter of fact, with DHA in an obstacle avoidance task the obstacle itself exerts a force on the stick which in turn makes the robot to change the movement direction even if the pilot is out of the loop. While, in the path following task of [79] (which according to the previous definitions would fall in the IHA class) when the stick moves on one side because of a future error in the path following, the error is doomed to rise if an external force (say the pilot) does not bring the stick in the center.

2.3 Reality-Based Haptic Aids

All the papers described so far are based on a haptic aid which does not exist in reality. In fact, they all artificially produce a haptic force linked to environmental constraints or to environmental goals (a specific target location, a path to follow or a desired maneuver).

One study [70] explores, instead, how to provide the UAV pilot with an enhanced indication about a real condition existing outside the aircraft: the authors, in fact, examine the value of haptic displays for alerting UAV operators about the onset of turbulence which is identified as being potentially detrimental to safe and effective UAV control by the UAV operators themselves. This is especially true for UAVs that require direct manual control in order to land.

The data in [70] reveal that haptic alerts, conveyed via the UAV operator's joystick, could indeed improve self-rated situation awareness during turbulent conditions in a simulated UAV approach and landing task. These improvements might result either from an increase in the operator's "presence" in the remote environment [77], from increased information by effective use of multi-sensory stimulation [86], or a combination of the two.

Before [70], turbulence was indicated solely by an unexpected perturbation of video images being transmitted from a UAV-mounted camera to the operator control station, appearing in the Head-Up Display (HUD).

Due to limitations inherent with reducing all environmental information to the visual channel, UAV operators may fail to perceive, or fail to correctly diagnose this video perturbation as sudden turbulences. In [70] visual feedback is supplemented by haptic feedback applied directly to the pilot's control stick, providing a redundant, kinesthetic alert: a force reflection in the axis-direction and scaled-ratio magnitude of the turbulence event.

In the same paper, four different alerts are evaluated and compared: Visual (perturbation of nose-camera imagery in the HUD - Baseline), Visual/Haptic (Visual and additional 1 second, low gain, high frequency vibration of the control stick), Visual/Aural (Visual and 1 second pure tone), Visual/Aural/Haptic (all three cues simultaneously). The data collected concern pilots performing simulated landing tasks. Conditions containing the haptic cue (Visual/Haptic and Visual/ Haptic/Aural) result in less error than non-haptic cue conditions (Visual and Visual/Aural). Although the aural alert also improves landing accuracy and detection of turbulence direction, performance is best with the redundant kinesthetic feedback. When randomly interrogated regarding the primary direction of the UAV immediately following a turbulence event, participants are more accurate when haptic feedback is present [70].

Interestingly, these results are true despite the haptic signals is not designed to closely simulate or mimic the veridical haptic information experienced by the pilot of a manned vehicle [53]. In fact, as said, the turbulence is transmitted through a low gain, high frequency vibration of the control stick.

2.4 Time Delays

As mentioned, a teleoperation system in presence of force feedback is referred as bilateral system. In such systems, the human operator controls a remotely located robot. The UAV operator is responsible for the UAV at all times, it is crucial that he/she at all times can understand the UAV. Informational transfers through the datalink have to be without delays that can have an effect on system performance and overall safety. It is vital that control inputs and orders can be executed immediately in emergency situations that require such actions. Datalink delays could be of various magnitude (from 100 to 1600 ms or more) and not always predictable to human operators, and can thus cause a lack of understanding with increased cognitive workload, decreased situational awareness and possible incorrect inputs as result with final failure of the mission [53].

The scientific literature suggests different ways to improve the performance of a teleoperated system in presence of time delay: starting from the *move and wait* strategy [75], that is initiating a control move and then waiting to see the response of the remote robot until the task is accomplished, to the more advanced control theory. The first methods regard automatic switching for stopping override [59], supervisory control [21] or the use of the predictive display [62, 37]. Beginning in the mid 1980s, more advanced control theoretic methods start to appear, such as Lyapunov-based analysis [49] and internal virtual model [23]. In the late 1980s and 1990s, network theory lead on to impedance representation [67] and passivity theory with [8, 52, 63] and without [11] the scattering variables (wave variables transformation). Reference [44, 27] through the two/four channel architectures and the impedance/hybrid matrix approach start mentioning the trade off between stability and transparency. In the 1990s the teleoperation through Internet begins and the problem of packets loss grows up [30]. Other methods overcoming the instability problems of bilateral teleoperation in presence of time delay are the admittance control [56, 57], the adaptive control [87] and the time domain passivity [27, 9]. Another way to handle the time delay communication and the loss of packets is the sampled Port-Hemiltonian approach [80].

In particular, while the passivity method presents a trade off between the stability and the transparency, the Port-Hemiltonian approach would allow both stable and transparent behavior [80].

3 Conventional Aircraft Artificial Feel

A typical trouble of remote piloting an RPV is the lack of situation awareness because of the physical separation between the pilot (inside the Ground Control Station, GCS) and the airborne RPV. Usually the GCS provides only the visual feedback; when an external disturbance or a fault, which on a conventional aircraft would produce a perceptible effect on the control column, affects the RPV, the pilot can understand it only by watching at the instruments. When a vertical wind gust disturbance affects a manned aircraft, the mutation of the angle of attack and wing load are almost instantaneous. This has also an immediate effect on mechanical-linkages and so on the control column. The altimeter on the GCS cockpit will show the resulting variation of altitude with a certain delay with respect to the actual disturbance time; as a matter of fact the aircraft dynamics from angle of attack to altitude has a low pass behavior and a phase lag (in the simplest linear approximation it behaves as an integrator). Thus mechanical controls also carry out the role of a tactile display: the human hand can interpret loading forces appearing on the handgrip in terms of demands imposed on the system and its expectable response, enabling the pilot to develop a beneficial phase lead [45].

As said in Section 1.4, usually the automation does not provide or could hinder SA if the designers fail in adequately addressing the operator need of it. Automation can also, in many different ways, be created to support good SA through decision aids and system interfaces. IHA-CAAF is introduced to satisfy such a different way to create SA.

Operators are prone in overlooking crucial information to sustain SA, though all relevant and needed informations are present. This was found to be the most frequent causal factor associated with SA errors [53]. Through the IHA-CAAF they do not have to think about the action to be taken in answer to the haptic stimulus because IHA-CAAF is built in order to produce a natural and instinctive response.

Furthermore, by considering that UAVs pilots are also manned aircrafts pilots, they expect, in presence of external disturbances such as wind gusts or turbulences, a stick cue which is similar to the one they would feel while piloting the aircraft on board. Thus, a good way to inform the remote pilot about the external disturbances could be perhaps to reproduce, through the haptic feedback, a feeling which mimics the real one.

The IHA-CAAF haptic feedback will be shown to increase the performance in terms of instinctive response to a stimulus in pilots without any previous training on the experiment (the author published the results in [2, 3, 4]). It also improves the situational awareness intended as making the pilot to feel as piloting the aircraft on board. This would improve the safety of the teleoperation by keeping higher the attention of the pilot in the task.

3.1 FBW Aircrafts/UAVs Analogy

As said this work is based on UAV feedback augmentation; nonetheless similar techniques could be employed in similar fields like Fly-By-Wire (FBW) piloted commercial aircrafts or helicopters.

A FBW system is an electrically-signaled aircraft control system (a computer-configured controller) that modifies the manual inputs of the pilot in accordance with control parameters. The displacements of the flight controller, the *sidestick*, are converted to electronic signals, and flight control computers determine how to move the actuators at each control surface to provide the expected response.

FBW aircrafts (Airbus, Boeing 777 and later designs) present, at least as concerning the haptic feedback, similar loss of situational awareness compared to the previous technology, i.e. the mechanically driven aircrafts (see later the Section 3.2).

In fact, FBW system employed both in large airliners and in military jet aircrafts, dispenses all the complexity of the mechanical circuit of the mechanical flight control system and replaces it with an electrical circuit. The FBW (also referred as *irreversible* control system [69]) makes use of an electronic passive sidestick in place of the conventional control stick which was connected to the actual aerodynamic surfaces via mechanical linkages (*reversible* control system [69]). The sidestick is in general implemented as a spring system with constant stiffness that makes the force felt by the pilot stronger as the displacement of the stick increases independently from the particular aerodynamic situation (velocity, load factor). Sometimes the sidestick may provide an artificial vibration (*stick shaker*) and some acoustical/visual warning that makes the pilot to know that the limits of the flight envelope (see Section A.1.1 for details) are going to be reached [81] or dead zones and discontinuities in the force/displacement characteristic.

The employment of fully powered controls made essential the introduction of completely artificial feel [25]. At that time, a considerable speculation about what elements of natural feel should be emulated, started. It was also coupled with the natural desire to minimize the cost and complexity of the feel devices.

The possibilities included control force variation with dynamic pressure (*q feel*), speed (*V feel*) or control deflection only (*spring feel*). Devices such as bobweights and downsprings which were already familiar on conventional aircraft, were sometimes included as well.

Artificial feel becomes more and more fundamental in addition to the visual cue in the context of RPVs.

3.2 Mechanically Driven Aircrafts

As said, a meaningful way to inform the remote pilot about the external disturbances is the reproduction, through the haptic feedback, of a feeling which mimics the one transmitted to the pilot on board of a manned mechanically driven aircraft. In this case, the pilot feels all the aerodynamic forces (external disturbances as wind gusts and turbulences) directly on the control column. The force felt by a pilot on the aircraft bar of a mechanical Flight Control System (FCS) during a maneuver depends in a very complex manner on all the aerodynamic characteristics of the aircraft, the current state of the aircraft (speed, angle of attack etc.) and of course on control device deflection. By taking into consideration the only longitudinal dynamics (pitch and altitude motion), the force felt by the pilot of a mechanically driven aircraft is [69]:

$$F_S = \eta_h C_h q S_e c_e G_e = (C_{h0} + C_{h,\alpha}\alpha_h + C_{h,\delta}\delta_e) \cdot \eta_h q S_e c_e G_e \qquad (3.1)$$

where η_h is the dynamic pressure ratio at horizontal tail, C_h is the elevator hinge moment, q is the dynamic pressure of the aircraft which is defined as

$$q = \frac{1}{2}\rho V^2$$

(where ρ is the air density and V is the airspeed), S_e and c_e are the elevator surface and chord and G_e is a gearing factor (with units) to convert moments to force and includes the geometry of the control mechanisms, pulleys, push-rods and cables (see Figure 3.1). C_{h0} is the elevator hinge moment coefficient at zero lift; $C_{h,\alpha}$ and $C_{h,\delta}$ are respectively the elevator hinge moment coefficient derivative with respect to tail angle of attack (α_h) and with respect to elevator deflection (δ_e).

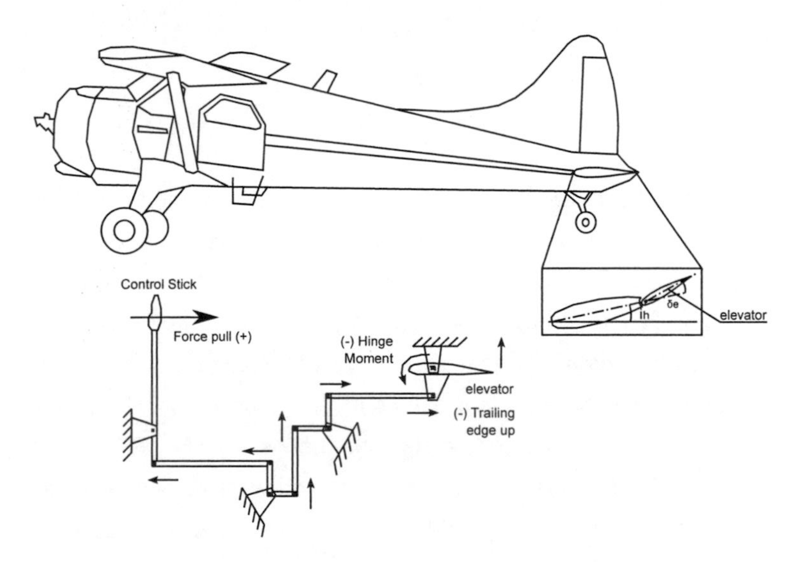

Figure 3.1. Mechanically driven aircraft [69]. i_h is the horizontal tail angle and δ_e is the elevator deflection.

A simplified expression for the force felt by the pilot of a mechanically driven aircraft can be re-written as in Section 3.2.1 (please find in Section 3.2.2 the mathematical proof).

3.2.1 A simplified stick force

A simplified expression for the force felt by the pilot of a mechanically driven aircraft can be re-written as made up, in general, by two different components: a spring-damper component, F_{SD}, and an external force component, F_{WG} (see Equation (3.2)).

$$F_S = F_{SD} + F_{WG} \tag{3.2}$$

where:

$$\begin{cases} F_{SD} = K \cdot \Delta\delta_e \\ K = \eta_h S_e c_e G_e |C_{h,\delta}| \cdot q \\ F_{WG} = \eta_h S_e c_e G_e |C_{h,\alpha}| \cdot q(\alpha - \alpha_{trim})(1 - \frac{d\varepsilon}{d\alpha}) \end{cases} \tag{3.3}$$

$\Delta\delta_e$ is the change in the commanded elevator deflection with respect to the trim condition deflection. α is the aircraft angle of attack, which is the angle between the direction of motion (relative velocity) and the x-axis of the Body Reference Frame (left-handed frame with origin in the center of gravity of the aircraft, $\mathbf{x_B}$ axis is in the vertical plane of symmetry of the aircraft and points the nose, $\mathbf{y_B}$ axis is in the plane perpendicular to the plane of vertical symmetry and points to the right side), α_{trim} is the angle of attack in trim condition (see later), ε is the downwash angle produced on the horizontal tail by the wings airflow. A justification for the approximate expression of Equation (3.2) and (3.3) is given in the Section 3.2.2.

3.2.2 Simplified Stick Force Proof

The longitudinal steady state equations of horizontal flight in Wind Axes (left-handed coordinate system with $\mathbf{x_W}$ same direction as the relative velocity and $\mathbf{z_W}$ downward, origin in the aircraft center of gravity) are written as [69]:

$$\begin{cases} W = L = C_L \cdot qS \\ 0 = m = C_m \cdot cqS \end{cases} \tag{3.4}$$

where W, L and m are respectively the aircraft total weight, lift and pitching moment; C_L and C_m are respectively the aircraft lift and pitching moment coefficients. c is the mean wing chord. The Equation (3.4) can be re-written as:

$$\begin{cases} mg = (C_{L0} + C_{L\alpha} \cdot \alpha + C_{L,ih} \cdot i_h + C_{L\delta} \cdot \delta_e) \cdot qS \\ 0 = (C_{m0} + C_{m\alpha} \cdot \alpha + C_{m,ih} \cdot i_h + C_{m\delta} \cdot \delta_e) \cdot qS \end{cases} \tag{3.5}$$

In Equation (3.5), C_{L0} and C_{m0} are respectively lift and pitching moment coefficients for zero angle of attack α. $C_{L\alpha}$, $C_{L,ih}$, $C_{L\delta}$ represent the change in lift coefficient with the angle of attack (the aircraft lift curve slope), α, the horizontal tail incidence angle, i_h, and the elevator deflection, δ_e (see Figure 3.1) respectively. $C_{m\alpha}$, $C_{m,ih}$ and $C_{m\delta}$ are equivalent variations of the pitching moment coefficient. As usual, q and S are dynamic pressure and wings area. The solutions of Equation (3.5) are referred as trim condition variables [69]:

$$\begin{cases} \alpha = \frac{(C_{L,trim} - C_{L0} - C_{L,ih} \cdot i_h) C_{m\delta} + (C_{m0} + C_{m,ih} \cdot i_h) C_{L\delta}}{(C_{L\alpha} C_{m\delta} - C_{m\alpha} C_{L\delta})} = \alpha_{trim} \\ \\ \delta_e = \frac{-C_{L\alpha}(C_{m0} + C_{m,ih} \cdot i_h) - C_{m\alpha}(C_{L,trim} - C_{L0} - C_{L,ih} \cdot i_h)}{(C_{L\alpha} C_{m\delta} - C_{m\alpha} C_{L\delta})} = \delta_{e,trim} \end{cases} \tag{3.6}$$

In general the following is held:

$$\alpha_h = \alpha \cdot (1 - \frac{d\varepsilon}{d\alpha}) + i_h - \varepsilon_0 \tag{3.7}$$

In Equation (3.7), the average downwash angle caused by the wings on the horizontal tail is often expressed [69] through

$$\varepsilon = \varepsilon_0 + \frac{d\varepsilon}{d\alpha} \cdot \alpha$$

where ε_0 is the down wash angle at zero airplane angle of attack and $\frac{d\varepsilon}{d\alpha}$ is the change of the downwash angle, ε, with the angle of attack, α.

The force F_S that the pilot applies on the bar should be equal to the hinge moment [69] written in Equation (3.1).

By supposing the aircraft to be provided with a *trimmable horizontal stabilizer* (THS) that is possible to position in order to make the force of Equation (3.1) null, i.e. $i_h = i_{h,trim}$ (by considering the Equation (3.7) into the Equation (3.1) and solving for $F_S = 0$):

$$\begin{cases} i_{h,trim} = -\frac{1}{C_{h,\alpha}} \left(C_{h0} + C_{h\alpha} \cdot \alpha_{trim} (1 - \frac{d\varepsilon}{d\alpha}) - C_{h,\alpha}\varepsilon_0 + C_{h,\delta}\delta_{e,trim} \right) \\ \\ F_S = 0 \end{cases} \tag{3.8}$$

If the aircraft is trimmed (THS deflected $i_{h,trim}$) and by considering that the pilot could move the bar through the application of the force ΔF_S and thus the elevator by $\Delta\delta_e$, it is possible to write:

$$\begin{cases} \alpha = \alpha_{trim} + \Delta\alpha \\ i_h = i_{h,trim} + \Delta i_h \\ \varepsilon_0 = const \\ \delta_e = \delta_{e,trim} + \Delta\delta_e \\ \alpha_h = \alpha_{h,trim} + \Delta\alpha_h \\ \alpha_{h,trim} = \alpha_{trim} \cdot (1 - \frac{d\varepsilon}{d\alpha}) + i_{h,trim} - \varepsilon_0 \end{cases} \tag{3.9}$$

By considering the Equation (3.7) and that the THS is deflected $i_{h,trim}$ and fixed to that value (then $\Delta i_h = 0$), it is possible to calculate $\Delta\alpha_h$:

$$\Delta\alpha_h = \Delta\alpha \cdot (1 - \frac{d\varepsilon}{d\alpha}) \tag{3.10}$$

The corresponding stick force change is obtained by substituting the previous ones in the Equation (3.1):

$$\Delta F_S = \eta_h q S_e c_e G_e \Big(C_{h,\alpha} \Delta\alpha \Big(1 - \frac{d\varepsilon}{d\alpha} \Big) + C_{h,\delta} \Delta\delta_e \Big) \tag{3.11}$$

The change in α, $\Delta\alpha$, produced by the change in δ_e, $\Delta\delta_e$, with respect to trim conditions, α_{trim} and $\delta_{e,trim}$, can be written as:

$$\Delta\alpha = \alpha - \alpha_{trim} \tag{3.12}$$

$$\Delta\delta_e = \delta_e - \delta_{e,trim} \tag{3.13}$$

The Equation (3.13) is obtained by supposing that the THS is fixed in the horizontal trim conditions ($i_h = i_{h,trim}$). As a consequence, the

Equation (3.11) can be simply written as:

$$F_S = K \cdot \Delta\delta_e + F_{WG} \tag{3.14}$$

Where:

$$\begin{cases} K = \eta_h S_e c_e G_e |C_{h,\delta}| \cdot q \\ F_{WG} = \eta_h S_e c_e G_e |C_{h,\alpha}| \cdot q(\alpha - \alpha_{trim})(1 - \frac{d\varepsilon}{d\alpha}) \end{cases} \tag{3.15}$$

In Equation (3.15), the dynamic pressure and the angle of attack are the only non-constant values. Thus, the simplified stick force equation above has basically two components: an elastic term which stiffness (K) varies with the dynamic pressure and an external component (F_{WG}) which varies with the dynamic pressure and the angle of attack.

3.3 CAAF

A pilot flying a mechanically steered aircraft feels on the stick the aerodynamic forces generated on the actual control surfaces. The simple fact that the pilot feels the load factor (ratio between lift and aircraft weight, see Section 3.3.1) helps him/her to avoid flight conditions which might be dangerous for the aircraft structure. As another simple example, stall may happen during a steep climb maneuver; while approaching the stall condition the stick becomes looser informing the pilot of the risk to lose aircraft control. Furthermore, external disturbances like wind gusts which may be very dangerous if not appropriately and suddenly compensated in a constrained mission environment (e.g., flight close to mountains or to bridges), would produce an immediate effect on the stick.
A pilot flying a remotely UAV cannot read on the GCS cockpit instruments

useful information like load factor, "distance" from stall and external disturbances, thus the Conventional Aircraft Artificial Feel (CAAF) haptic aiding scheme is designed for the purpose of providing the pilot with a richer information than the visual display only. The experiments are performed in order to show and assess analytically that these additional haptic informations help the pilot from a performance point of view.

Level 1 SA (see Section 1.4) says that the pilot needs to accurately perceive information about the weather among other elements. Reference [70] follows this principle by creating a haptic sensation linked to the turbulence even if in such a case the haptic signal was not related to the real sensation experienced by a pilot of a manned aircraft (see Section 2.3). The present work instead introduces a haptic feedback which mimics aerodynamic forces usually experienced by the pilots of manned aircrafts and it belongs by definition to the class of IHA being born, above all, to improve the SA and not designed taking into account the right maneuver to perform in order to reject the wind gust.

As mentioned before, the newly introduced haptic feedback is given the name of *Conventional Aircraft Artificial Feel* (CAAF).

Two different versions of the CAAF are presented: the former, named *Variable Stiffness CAAF*, estimates the effect of wind gust as changes in stick stiffness K, while the external force, F_{WG}, is set to zero (see Section 3.3.1); the latter, named *Force Injection CAAF*, estimates the effect of wind gust as changes in the external force F_{WG} through the variations of angle of attack, α, and dynamic pressure, q, while the stick stiffness K is set to a constant value (see Section 3.3.2).

3.3.1 Variable Stiffness CAAF

The Variable Stiffness CAAF estimates the effect of wind gust as changes in stick stiffness according to a weighted sum of load factor, n, and dynamic pressure, q. Thus, the force is assumed to be dependent on the two most important variables for defining the flight envelope (see Section A.1.1 for details). The load factor

$$n = \frac{L}{W}$$

is defined as the ratio of the lift L to the weight W of the aircraft, thus it is a measure of the severity of a commanded maneuver. It is introduced in the stick force equation to make, as said, the pilot more conscious about the commanded maneuver by making more difficult the maneuvers which might be dangerous for the aircraft structure and cause accidents as the loss of wings in the RPV. The external force is set to zero:

$$\begin{cases} F_{CAAF,vs} = F_{SD,vs} + F_{WG,vs} \\ F_{SD,vs} = K_{S,vs} \cdot \delta_S + K_{D,vs} \cdot \dot{\delta}_S \\ F_{WG,vs} = 0 \end{cases} \tag{3.16}$$

$F_{SD,vs}$ is the Spring-Damper component of the force and $F_{WG,vs}$ is the external force component (vs is for *variable stiffness*). The Variable Stiffness CAAF, Equation (3.16), is similar to the Equation (3.14) except for the null external force component, for the introduction of the load factor in the variable stiffness and for the addition of a damper component as well in order to provide some damping for the future implementation of the CAAF in a haptic device. Being $\Delta\delta_e$ in Equation (3.14) the elevator deflection around the trim value, which is $0deg$ with the THS deflected i_{trim}, and fixed on this value and since the elevator deflection is proportional to the bar deflection for mechanically driven aircrafts, in Equation (3.16) the stick deflection δ_S is employed instead of $\Delta\delta_e$. $F_{CAAF,vs}$ represents the change in the stick force during a maneuver with respect to the stick force in trim conditions ($F_{trim} = 0$). δ_S and $\dot{\delta}_S$ are stick deflection and stick deflection rate respectively. $K_{D,vs}$ is the damping constant. Equation (3.17) shows the value of the stiffness in the Variable Stiffness CAAF.

$$K_{S,vs} = K_{f,vs} \cdot [K_{q,vs} \cdot q + K_n \cdot (n-1)] \tag{3.17}$$

It shows the changes of the stiffness as proportional to the squared velocity, through q, and to the load factor.

$K_{q,vs}$ and K_n are the weights of the dynamic pressure and of the difference between the load factor during the maneuver and the one in horizontal flight respectively $(n-1)$; $K_{f,vs}$ is a constant gain which determines the "amount" of force feedback.

The sign convention is the same as in [69] (see Figure 3.1). The force the pilot feels on the stick has the same sign as the deflection requested to the elevator: for example during a climb maneuver the pilot exerts a positive force on the stick by pulling it and feels a negative force that tries to push it back; while climbing the elevator is deflected upwards (negative angle). Thus in Equation (3.16), a positive value is needed as $K_{S,vs}$ and by watching at Equation (3.17) the following points must be satisfied:

- Being the dynamic pressure q positive by definition (see Section 3.2) because all the terms inside are positive in sign (the air density, ρ, and the squared velocity, V^2), a positive value is needed as $K_{q,vs}$. The goal of the dynamic pressure component is to make the pilot conscious about the velocity of the UAV: the higher is the velocity, the bigger is the dynamic pressure component, the bigger is the spring component and more difficult will be to perform a maneuver.
- Being the load factor positive for climbing maneuver and negative for diving maneuver and needing a positive sign of the product

$K_n \cdot (n-1)$, K_n should have a negative value for diving maneuvers and a positive value for climbing maneuvers.
The goal of the introduction of the load factor in the spring component of the Variable Stiffness CAAF is to avoid the pilot to do a sudden maneuver: the higher is the absolute value of the load factor, the bigger is the stiffness of the stick and more difficult will be to perform a maneuver.

Furthermore, $K_{q,vs}$ and K_n can be interpreted as the strain the pilot must exert on the bar to produce a change in velocity or a change in the load factor during a maneuver. In literature [69], something similar to K_n is referred as *stick-force-per-g*.

In order to assign meaningful values to the constants $K_{q,vs}$, K_n and $K_{f,vs}$, the dynamic pressure and the load factor components are normalized with respect to the maximum values they can assume. The choice made in Equation (3.18) (for the sake of simplicity only the right side up flight is considered) would satisfy the previous hypothesis:

$$\begin{cases} K_{q,vs} = \frac{K'_{q,vs}}{\frac{1}{2}\rho V_{max}^2} \geq 0, & V_{max} = V_D \\ K_n = \begin{cases} \frac{K'_n}{(n_1-1)} \geq 0, & for \quad n \geq 1 \Rightarrow K_n(n-1) \geq 0 \\ \frac{K'_n}{(n_2-1)} < 0, & for \quad n < 1 \Rightarrow K_n(n-1) > 0 \end{cases} \end{cases} \tag{3.18}$$

V_D is the design diving speed that, being unknown, is hypothesized to be the 110% of the never exceed speed, V_{NE} (see Section A.1.1). n_1 and n_2 are respectively the positive and negative maximum values of load factor of the aircraft (see Section A.1.1 for details).

As concerning K'_n and $K'_{q,vs}$, it could be interesting to find out the optimal values capable of minimizing a performance index. The value 0.5 for both of them is the first heuristic choice in this work. Being the constants normalized with respect to the maximum values of the variable they weight (q and n), then the value 0.5 means that the *feel* in Equation (3.17) is made up by the changes in q for the 50%, by the changes in n for the remaining 50%. The quantity in squared parenthesis in Equation (3.17) will assume the value 1 at maximum. As said, the amount of the feedback force depends on $K_{f,vs}$ which scales the stiffness to the desired value. The Federal Aviation Regulation (FAR) of the Federal Aviation Administration (FAA) and in particular the FAR 23 Sect.23.155 imposes the strength limits necessary to control the elevator for certain values of the load factor, but the real amount of force to employ will depend at the end on the haptic device maximum output force.

The final expression of the haptic feedback force becomes then:

$$\begin{cases} F_{CAAF,vs} = F_{SD,vs} + F_{WG,vs} \\ F_{SD,vs} = K_{S,vs} \cdot \delta_S + K_{D,vs} \cdot \dot{\delta}_S \\ K_{S,vs} = K_{f,vs} \cdot [K_{q,vs} \cdot q + K_n \cdot (n-1)] \\ F_{WG,vs} = 0 \end{cases} \tag{3.19}$$

The haptic feedback expression of Equation (3.19) is named **Variable Stiffness Conventional** (for mechanically-driven) **Aircraft Artificial Feel (CAAF)** due to its aerodynamically inspired nature. This type of force feedback, in analogy to what found in the artificial feel literature [25], could be addressed as a *qn-feel* system since the force it generates is proportional to both dynamic pressure (q) and load factor (n). This force is tested through the *CAAF Experiment* (see Section 3.6.1).

3.3.2 Force Injection CAAF

The Force Injection CAAF of Equation (3.20) estimates the effect of wind gust as changes in angle of attack α and dynamic pressure q and produces external force variations as opposed to the former version of Equation (3.19) that uses stick stiffness modifications. In the altitude regulation task (object of the first experiments) it was noticed that the velocity was close to the trim speed, V_{trim}, and the load factor to the unity as in horizontal flight ($n = 1$), then a constant value ($K_{S,fi}$) is chosen as stiffness while an external force component F_{WG} as in Equation (3.15) is considered:

$$\begin{cases} F_{CAAF,fi} = F_{SD,fi} + F_{WG,fi} \\ F_{SD,fi} = K_{S,fi} \cdot \delta_S + K_{D,fi} \cdot \dot{\delta}_S \\ K_{S,fi} = K_{f,fi} \cdot K_{q,fi} \cdot q_{trim} \\ F_{WG,fi} = \eta_h S_e c_e G_e |C_{h,\alpha}| \cdot q(\alpha - \alpha_{trim})(1 - \frac{d\varepsilon}{d\alpha}) \end{cases} \tag{3.20}$$

$F_{SD,fi}$ is the Spring-Damper component of the force and $F_{WG,fi}$ is the external force component (fi is for *force injection*). As previously, a damper component with damping constant $K_{D,fi}$ is added as well in order to provide some damping for the future implementation of the CAAF in a haptic device. q_{trim} is the dynamic pressure related to the trim velocity, V_{trim}.

$F_{CAAF,fi}$ represents the change in the stick force during a sudden vertical wind gust. The wind gust affects the angle of attack and moves it away from the one in trim conditions, α_{trim}. δ_S and $\dot{\delta}_S$ are the stick deflection and the stick deflection rate respectively.

$K_{q,fi}$ and $K_{f,fi}$ are respectively the weight of the dynamic pressure in trim conditions and a constant gain which determines the "amount" of Spring-Damper force $F_{SD,fi}$. In the following, the external force $F_{WG,fi}$ will be indicated as $F_{WG,IHA}$. Concerning the meaning of variables present in Equation (3.20), please refer to Section 3.2.

The Equation (3.20) can be written as:

$$\begin{cases} F_{CAAF,fi} = F_{SD,fi} + F_{WG,IHA} \\ F_{SD,fi} = K_{S,fi} \cdot \delta_S + K_{D,fi} \cdot \dot{\delta}_S \\ K_{S,fi} = K_{f,fi} \cdot [K_{q,fi} \cdot q_{trim}] \\ F_{WG,IHA} = K_{fWG,fi} \cdot [K_{q,\alpha} \cdot q(\alpha - \alpha_{trim})] \end{cases} \tag{3.21}$$

$K_{q,\alpha}$ and $K_{fWG,fi}$ are respectively the weight of $q(\alpha - \alpha_{trim})$ and a constant gain which determines the "amount" of external force $F_{WG,IHA}$.

As concerning the sign, the force the pilot feels on the stick during a vertical wind gust has the same sign as the deflection caused to the elevator by the wind gust. For example a downward wind gust (negative) will create a positive elevator deflection (trailing edge down), a drop in the angle of attack ($\alpha - \alpha_{trim} < 0$) and so a negative stick deflection δ_S (i.e. forwards; see Figure 3.2). Consequently, the force felt by the pilot is negative (the bar tends to move away from the pilot) for downwards (negative) wing gusts, while it is positive (the bar tends to move towards the pilot) for upward (positive) wind gusts. Hence, a positive value as $K_{S,fi}$ and a force $F_{WG,IHA}$ concordant with $(\alpha - \alpha_{trim})$ are needed and by watching at Equation (3.21), the following points must be satisfied:

- Being the dynamic pressure q positive by definition (see Section 3.2) because all the terms inside are positive in sign (the air density, ρ, and the squared velocity, V^2), a positive value is needed as $K_{q,fi}$.

- Being the dynamic pressure q positive (see above) and being the change in angle in attack $(\alpha - \alpha_{trim})$ positive for $(\alpha > \alpha_{trim})$ and negative for $(\alpha < \alpha_{trim})$ and needing an external force $F_{WG,IHA}$ concordant with $(\alpha - \alpha_{trim})$, $K_{q,\alpha}$ should have a positive value.

The goal of spring and external force components is to make the pilot conscious about the change in velocity and angle of attack of the UAV produced by the wind gust: a downward wind gust produces, as said, a diving maneuver (the velocity grows up and the angle of attack falls off) and the haptic feel in Equation (3.20) would suggest that the aircraft is diving and a pilot input in the opposite direction (i.e. pulling the bar) is needed in order to restore the previous trim condition value. The stronger is the gust, the bigger is the change in angle of attack and in the velocity produced, the bigger are the spring-damper and the external force components and a stronger and clearer information about the presence of a wind gust will be given to the pilot. An improvement of the situational awareness about the external conditions of the aircraft will be produced. As said, the action requested to the pilot in order to restore the previous trim conditions is to counteract the haptic feel. This would be a natural reaction to the force for what Schmidt and Lee proved [73] (see Section 2.2).

In order to assign meaningful values to the constants $K_{q,fi}$, $K_{q,\alpha}$, $K_{f,fi}$ and $K_{fWG,fi}$, the dynamic pressure in trim conditions, q_{trim}, and the product $q(\alpha - \alpha_{trim})$ are normalized with respect to the maximum values they can assume. The choice made in Equation (3.22) (for the sake of simplicity only the right side up flight is considered) would satisfy the previous hypothesis:

$$\begin{cases} K_{q,fi} = \dfrac{K'_{q,fi}}{\frac{1}{2}\rho V_{max}^2} \geq 0, & V_{max} = V_D \\ K_{q,\alpha} = \begin{cases} \dfrac{K'_{q,\alpha}}{\frac{1}{2}\rho V_{max}^2 \cdot (\alpha_{S1} - \alpha_{trim})} \geq 0, & for\ \alpha \geq \alpha_{trim} \Rightarrow \\ & \Rightarrow K_{q,\alpha} \cdot q(\alpha - \alpha_{trim}) \geq 0 \\ \dfrac{K'_{q,\alpha}}{\frac{1}{2}\rho V_{max}^2 \cdot (\alpha_{S1} - \alpha_{trim})} \geq 0, & for\ \alpha < \alpha_{trim} \Rightarrow \\ & \Rightarrow K_{q,\alpha} \cdot q(\alpha - \alpha_{trim}) < 0 \end{cases} \end{cases} \tag{3.22}$$

Furthermore, $K_{q,fi}$ and $K_{q,\alpha}$ can be interpreted as the strain the pilot must exert on the bar to produce a change in velocity and a change in the angle of attack a maneuver.

$V_{max} = V_D$ where V_D is the design diving speed that, being unknown, is hypothesized to be the 110% of the never exceed speed, V_{NE} (see Section A.1.1).

α_{S1} and α_{trim} are the right side up angle of attack respectively in stall and trim conditions (see Section A.1.1 for details).

As concerning $K'_{q,fi}$ and $K'_{q,\alpha}$, it could be interesting to find out the optimal values capable to minimize a performance index. The first heuristic choice in this work was the value 0.5 for both of them.

Being the constants normalized with respect to the maximum values of the variable they weight (q_{trim} and $q \cdot (\alpha_{S1} - \alpha_{trim})$), then the value 0.5 for both of them means that their values are the 50% of the maximum available ones. The quantities in squared parenthesis in Equation (3.21) will assume both the value 0.5 at maximum. The amount of stiffness and the amount of the external force strongly depend on $K_{f,fi}$ and $K_{fWG,fi}$ respectively. They scale the stiffness and the external force $F_{WG,IHA}$ to the desired value. Their choice is made heuristically by taking into account the haptic device maximum output force.

The final expression of the haptic feedback force is represented by the Equation (3.21) and is named **Force Injection Conventional** (for mechanically-driven) **Aircraft Artificial Feel (CAAF)** due to its aerodynamically inspired nature. This type of force feedback, in analogy to what found in the artificial feel literature [25], could be addressed as a $q\alpha - feel$ system since the force it generates is proportional to both dynamic pressure (q) and angle of attack (α). This force is tested through the *CAAF VS DHA Experiment* (see Section 3.6.3).

Dickinson noted that "we can take the opportunity of making control forces do what we desire them to do rather than having to accept the consequences of fundamental laws as hitherto" [25]. Thus from now on, the mentioned opportunity is taken by using heuristic stiffness, damping constants and external forces instead of using constants (as in Equations (3.18) and (3.22)) which depend on the particular aircraft under consideration. This would make the haptic force to be transportable because created on the human being feeling instead of the particular aircraft (remotely or not) piloted.

3.4 The Experimental Setup

In order to test the CAAF concepts exposed in Sections 3.3.1 and 3.3.2, a simulated flight experiment is set-up. A fully non linear aircraft simulator is used to provide a realistic aircraft response. An aircraft simulator is implemented using a Matlab/Simulink simulation. The selected aircraft model is a De Havilland Canada DHC-2 Beaver implemented using the Flight Dynamics and Control Toolbox [68].

The selected haptic device is the widely used Omega Device in Figure 3.2 (omega.3, Force Dimension, Switzerland) which is chosen in order to simulate a control column of a mechanically driven aircraft. It is a 3DOF high precision force feedback device which provides control stick simulated force up to $12N$ (See Section A.2 for details).

A simulated Electronic Flight Instrument System display (EFIS display, Figure 3.3) is used during the experiments to produce the visual cues. It was designed to be as similar as possible to conventional aircraft head-down display (see Section A.3 for details on the EFIS display implementation). The display shows the relevant variables in the task (pitch, altitude, speed) and the variable to be regulated (altitude) with a

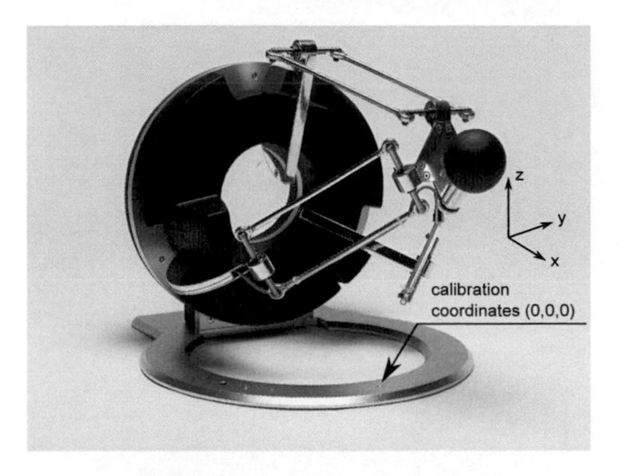

Figure 3.2. The Omega Device with reference frame.

magenta reference mark for the set point $300ft$ for altitude.

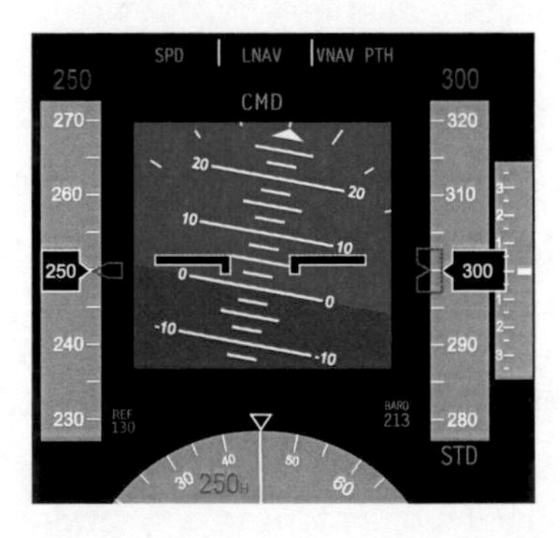

Figure 3.3. The Electronic Flight Instrument System display.

Figure 3.4 shows the experimental test bed comprising of a video display and the haptic device.

The only dynamics being considered in this Chapter is the longitudinal one. In order to control the longitudinal dynamics, the pilot usually acts on the thrust and on the elevator. In the present work, the elevator deflection is, by hypothesis, the only input provided to the simulated aircraft. This is a reasonable choice being the present work an artificial feel study. In fact,

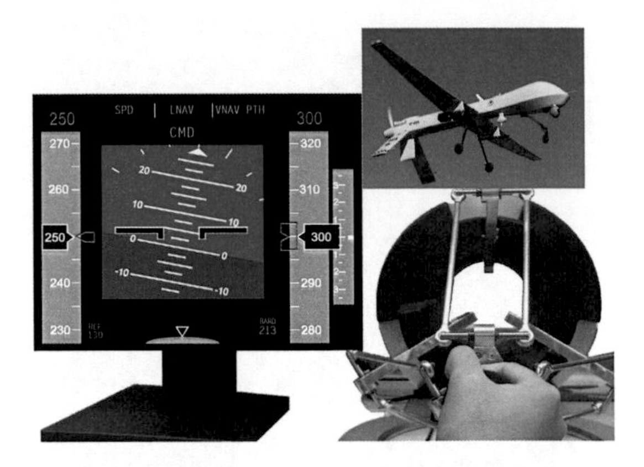

Figure 3.4. The wind gust rejection experimental setup.

acting on thrust and on the elevator at the same time would be reasonable for an autopilot or a Stability Augmentation System (SAS) study. Acting on thrust and on the elevator at the same time is also usually useless or undesirable, even on a real aircrafts (i.e. during the takeoff in which pulling-up the aircraft through the elevator supplying the maximum thrust is needed). Furthermore, acting only on the elevator to pull-up the aircraft is a traditional piloting maneuver.

In this work, the elevator deflection is proportional to the stick deflection δ_S of Equations (3.19) and (3.21). δ_S is the input to the aircraft and can be thought as generated by the operator through moving the Haptic Device end-effector in the $x-direction$ (see Figure 3.2).

An artificial impulse (simulating a vertical wind gust) on the elevator, starts the natural longitudinal aircraft modes: the Phugoid and the Short Period modes (see Section A.1.2). It causes a dynamic transient phase because of the exchanges between kinetic and potential energy and oscillations in the aircraft longitudinal variables (velocity, pitch angle, altitude, etc) around the center of gravity start. In Figure 3.5 the mentioned natural modes are shown (black line): the Phugoid is the most visible oscillation, while the Short Period (characterized by complex and conjugate poles) oscillation has, as the name suggests, a short period and, since it has usually a big damping constant, it disappears very soon.

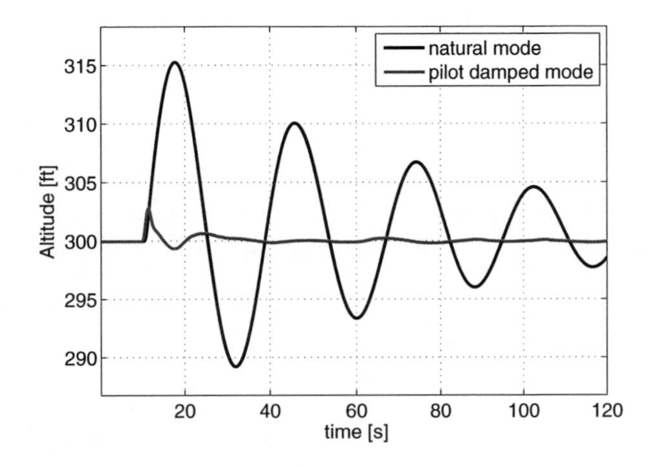

Figure 3.5. Response to elevator impulse input: Phugoid and Short Period natural aircraft modes (black line) versus the typical aircraft response damped by a good pilot (red line).

The Phugoid mode is characterized as well by complex and conjugate poles that produce a lightly damped oscillation during which the dynamic pressure, the wing load factor and the aircraft angle of attack change because of the changes in the aerodynamic forces acting on the aircraft. The pilot (or the autopilot) is needed to extinguish them with stick commands by holding the pitch angle through the use of the *artificial horizon*. Figure 3.5 shows as well (red line) a sample time history when a pilot acts on the stick to regulate it.

Since the subjects have to control the longitudinal dynamics only, the haptic aid for the wind gust rejection task is set only in the control device longitudinal axis (x axis in Figure 3.2).

The general force expression employed in both the just mentioned disturbance rejection experiments is given in Equation (3.23):

$$\begin{cases} F_{S,x} = F_{SD,x} + F_{WG,x} \\ F_{SD,x} = F_{SD} = F_{S,x} + F_{D,x} \\ F_{WG,x} = F_{WG} \end{cases} \tag{3.23}$$

In Equation (3.23), F_{SD} and F_{WG} indicate the Spring-Damper force and the external force of either the Equation (3.19) ($F_{SD,vs}$ and $F_{WG,vs}$) or the Equation (3.21) ($F_{SD,fi}$ and $F_{WG,IHA}$). Remember that $F_{WG,IHA}$ of Equation (3.21) corresponds to $F_{WG,fi}$ of Equation (3.20).

$$F_x = K_{S,x} \cdot x_S + K_{D,x} \cdot \dot{x}_S + F_{WG} \tag{3.24}$$

The force F_x felt by the operator during the wind gust rejection task (see Equation (3.23) and (3.24)) along the control device x axis (see Figure 3.2) is a combination of the elastic term $F_{S,x}$, $(K_{S,x} \cdot x_S)$ with constant stiffness $K_{S,x}$, the damping term $F_{D,x}$, $(K_{D,x} \cdot \dot{x}_S)$ with a damping constant $K_{D,x}$ (refer to the Table A.2 for the values used), and an external force component F_{WG}. x_S and $\dot{x}_S$ are the longitudinal displacement and displacement rate of the end-effector respectively, in the following indicated as δ_S and $\dot{\delta}_S$.

3.5 Disturbance Rejection Experiments

Two experiments run within the specific field of Remotely Piloted Vehicles control in a disturbance rejection task are described: the CAAF Experiment and the CAAF VS DHA Experiment.

The aim of the **CAAF Experiment** is to prove the effectiveness of the newly developed IHA-Variable Stiffness CAAF with respect to the absence of force feedback (NoF with only visual feedback and gravity compensation on the control device). See Section 3.6.1 for details.

The aim of the **CAAF VS DHA Experiment** is to compare three approaches:

- the newly developed and just described IHA-based Force Injection CAAF;
- the DHA force (see later);
- the NoEF, a force which is only linked to the actual displacement of the control device.

Sections 3.5.1 and 3.5.2 describe the simulators built in order to run the above mentioned experiments.

3.5.1 The CAAF Experiment Simulators

NoF Simulator

Figure 3.6 shows the block diagram of the simulation system used to test the NoF feedback. The altitude error (between desired altitude H_t and aircraft altitude H), e_H, is fed to the pilot P via the visual display showing the altitude error (see Figure 3.3). The pilot force input (F_h), is fed to the haptic device (OD block in Figure 3.6) to produce the stick deflection δ_S (used, in absence of wind gusts, directly as aircraft elevator control input by hypothesis). δ_{WG}, representing the wind gust disturbance, is summed up to the stick deflection to produce the elevator input to the aircraft δ_e.

Under the NoF condition no haptic feedback is transmitted to the pilot:

$$F_{NoF} = 0 \tag{3.25}$$

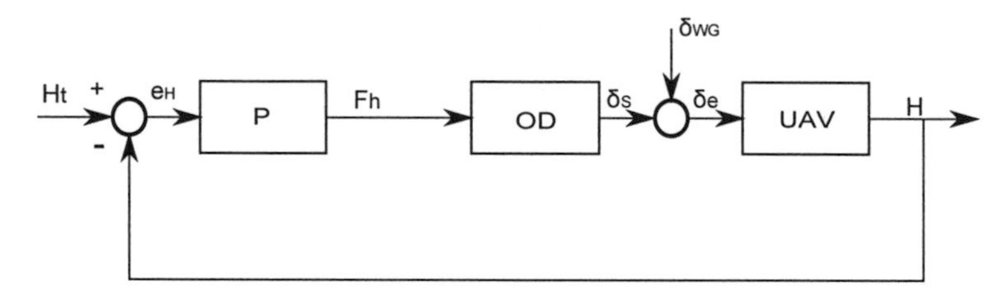

Figure 3.6. NoF simulator scheme.

In fact, the NoF condition represents a condition in which neither the elastic or damping forces are fed-back to the pilot. Not even the gravity force is transmitted to the pilot being the gravity compensation is activated in the haptic device.

Suppose a wind gust affects the aircraft: being both $F_{SD,x}, F_{WG} = 0$ in the Equation (3.23), no force is directly linked either to the wind gust or to the actual end-effector displacement. Thus, the pilot will not feel through the sense of touch any haptic information about both the position of the control device end-effector and the presence of wind gust; he/she will just see how the altitude evolves through the visual EFIS display (see Figure 3.3). The visual feedback is the same in all the conditions of the experiment.

IHA-Variable Stiffness CAAF Simulator

Figure 3.7 shows the block diagram of the simulation system used to test the IHA-Variable Stiffness CAAF feedback. The altitude error (between desired altitude H_t and aircraft altitude H), e_H, is fed to the pilot P via the visual display showing the altitude error (see Figure 3.3). The pilot force input (F_h), is fed to the haptic device (OD block in Figure 3.7) to produce the stick deflection δ_S (used, in absence of wind gusts, directly as aircraft elevator control input by hypothesis) and its rate of change $\dot{\delta}_S$. δ_S and $\dot{\delta}_S$ feedback indicate the proprioceptive feedback. δ_{WG}, representing the wind gust disturbance, is summed up to the stick deflection to produce the elevator input to the aircraft δ_e. The load factor n and the velocity V are employed to calculate the haptic feedback of Equation (3.19).

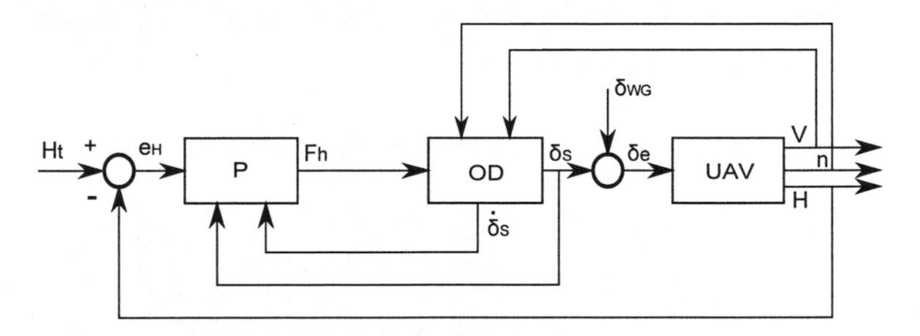

Figure 3.7. IHA-Variable Stiffness CAAF simulator scheme.

Suppose a wind gust affects the aircraft: the pilot, while damping the Phugoid mode, will feel a force feedback proportional to the changes in the dynamic pressure and in the load factor according to the Equation (3.19) and will get the same visual feedback as in the NoF condition.

3.5.2 The CAAF VS DHA Experiment Simulators

IHA-Force Injection CAAF Simulator

Figure 3.8 shows the block diagram of the simulation system used to test the IHA-Force Injection CAAF feedback. The altitude error (between desired altitude H_t and aircraft altitude H), e_H, is fed to the pilot P via the visual display showing the aircraft speed and altitude (see Figure 3.3). The aircraft speed (V), used to compute the dynamic pressure, and the angle of attack (α) are fed to the Haptic device that implements the CAAF-IHA law and feeds-back the force ($F_{WG,IHA}$) as in Equation (3.21) which, together with the pilot force input (F_h), is fed to the haptic device (OD block in Figure 3.8) to produce the stick deflection δ_S (used, in absence of wind gusts, directly as aircraft elevator control input by hypothesis) and its rate of change $\dot{\delta}_S$. δ_S and $\dot{\delta}_S$ feedback indicate the proprioceptive feedback. δ_{WG}, representing the wind gust disturbance, is summed up to the stick deflection to produce the elevator input to the aircraft δ_e. The angle of attack α and the velocity V are employed to calculate the haptic feedback according to the Equation (3.21).

The force felt in IHA case can be summarized as in the Equation (3.26):

$$F_{IHA} = F_{SD,x} + F_{WG,IHA} \tag{3.26}$$

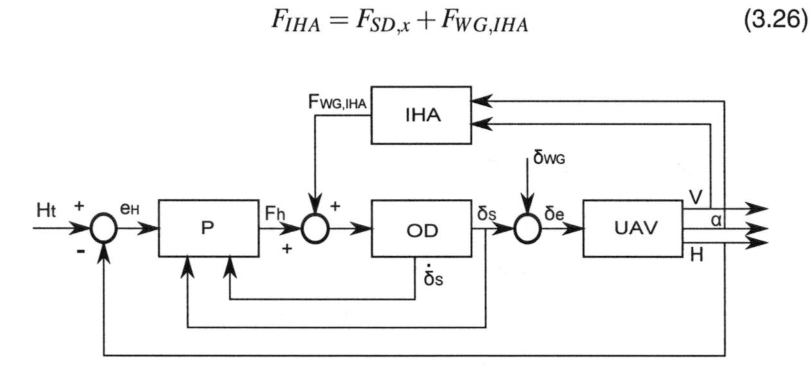

Figure 3.8. IHA-Force Injection CAAF simulator scheme.

Suppose a downward wind gust affects the aircraft: the angle of attack of the aircraft decreases with respect to the trim condition, the dynamic pressure changes (possibly very lightly depending on the gust speed with respect to the aircraft speed) and the altitude tends to decrease. Within this condition, the CAAF-IHA law produces a negative force, $F_{WG,IHA}$, that would produce a negative stick deflection, δ_S, and thus induces the aircraft to dive even more. The force is immediately felt by the pilot who learns that something changed. In this specific case the pilot feels a force that pulls the stick, that is to dive, and he/she should react immediately, according to his/her experience, by opposing to the stick motion in order to keep the altitude constant. This type of force feedback, roughly speaking with opposite sign than the actual maneuver to be taken, is in complete accordance with the IHA concept.

Figure 3.9 depicts an example of the variables history during a simulation trial.

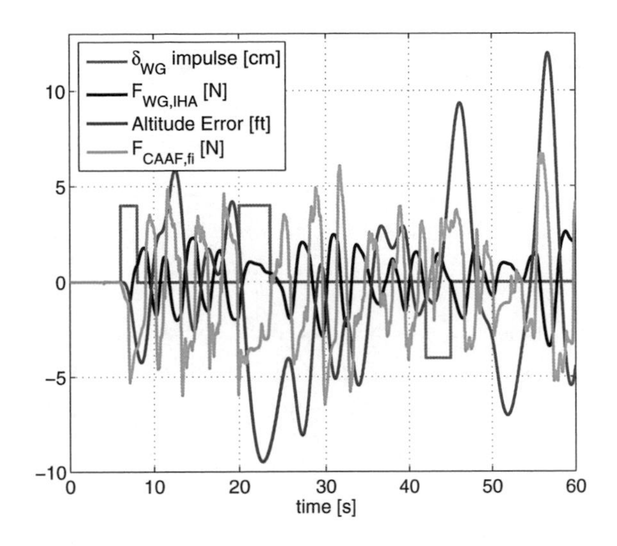

Figure 3.9. IHA-Force Injection CAAF simulation example.

NoEF Simulator

Figure 3.10 shows the block diagram of the simulation system used to test the NoEF feedback. The altitude error (between desired altitude H_t and aircraft altitude H), e_H, is fed to the pilot P via the visual display showing the aircraft speed and altitude (see Figure 3.3). The pilot force input (F_h), is fed to the haptic device (OD block in Figure 3.10) to produce the stick deflection δ_S (used, in absence of wind gusts, directly as aircraft elevator control input by hypothesis) and its rate of change $\dot{\delta}_S$. δ_S and $\dot{\delta}_S$ feedback indicate the proprioceptive feedback. δ_{WG}, representing the wind gust disturbance, is summed up to the stick deflection to produce the elevator input to the aircraft δ_e.

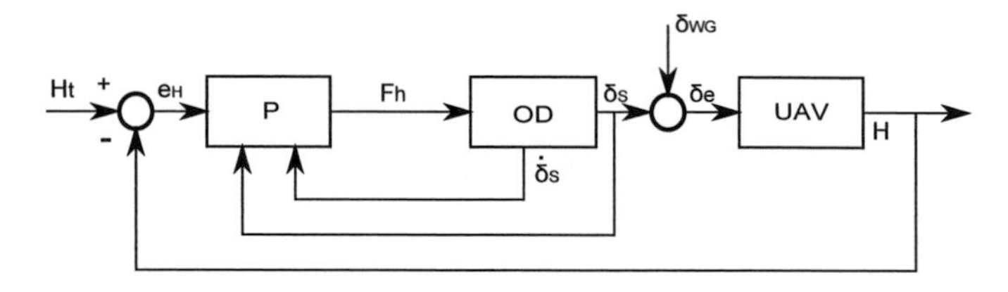

Figure 3.10. NoEF simulator scheme.

The NoEF condition presents a constant stiffness stick ($K_{S,x}$ in Table A.2) and simulates a fly-by-wire like situation. In the NoEF condition the force exerted by the haptic device is the same (i.e. the same constant Spring-Damper component) as in the Equation (3.21) except for $F_{WG,IHA}$ which is set to zero in this condition. The pilot has the EFIS display (see Figure 3.3) as the only instrument showing the aircraft speed and altitude. The visual feedback is the same in all the conditions of the experiment.

Under the NoEF condition, the haptic feedback of Equation (3.27) is transmitted to the pilot.

$$F_{NoEF} = F_{SD,x} \tag{3.27}$$

Suppose a wind gust affects the aircraft: missing in Equation (3.27) any term linked to the wind gust, the pilot will not feel any haptic information about the possible presence of it; he/she will just see the variation of altitude only through the visual display. The only haptic feedback felt by the pilot is proportional to δ_S and $\dot{\delta}_S$ produced only by the own input force F_h.

Figure 3.11 depicts an example of the variables history during a simulation trial.

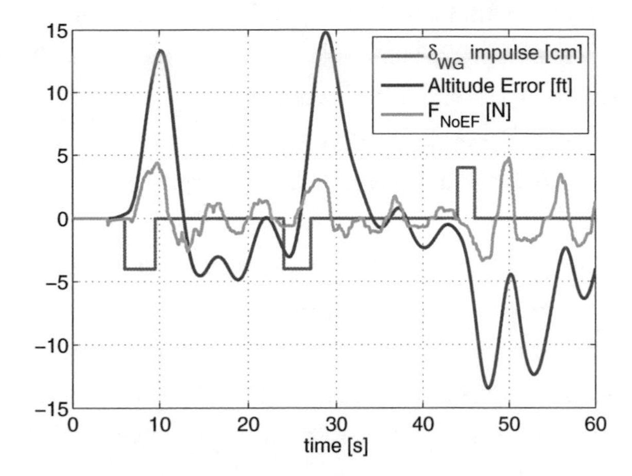

Figure 3.11. NoEF simulation example. F_{WG} (not shown) is null in this case.

Compensator-Based DHA Simulator

In order to compare the three approaches, a DHA-based simulator is designed. According to the DHA definition, a Direct Haptic Aiding system for wind gust rejection should produce a force or a change in stiffness that helps the pilot "directly" in achieving the task that is in this case to reject the gust. Thus, a system that produces a force which moves the stick in the same direction the pilot should do to reject the disturbance, seems appropriate for a DHA control. As a matter of fact, the obstacle avoidance system described in [40, 43] works exactly according to such a principle: stiffness variation together with force feedback were investigated and found to be able to provide better results than single stiffness or force feedback [43]. Nevertheless, for the purposes of this comparison, force feedback only is investigated and compared. A compensator is added to compute the external force to be felt by the pilot. The Haptic device is controlled to behave as a spring-damper system (same spring-damper constants as in IHA and NoEF cases) with an additional force $F_{WG,DHA}$ which is generated by the DHA compensator (see later).

Figure 3.12 shows the block diagram of the simulation system used to test the DHA concept. The altitude error (between desired altitude H_t and aircraft altitude H), e_H, is fed to the pilot P via the visual display showing the aircraft speed and altitude (see Figure 3.3). The altitude error, e_H, is also fed to the *DHA* block that implements the DHA force and feeds-back the force ($F_{WG,DHA}$) which, together with the pilot force input (F_h), is fed to the haptic device (OD block in Figure 3.12) to produce the stick deflection δ_S (used, in absence of wind gusts, directly as aircraft elevator control input by hypothesis) and its rate of change $\dot{\delta}_S$. δ_S and $\dot{\delta}_S$ feedback indicate the proprioceptive feedback. δ_{WG}, representing the wind gust disturbance, is summed up to the stick deflection to produce the elevator input to the aircraft δ_e.

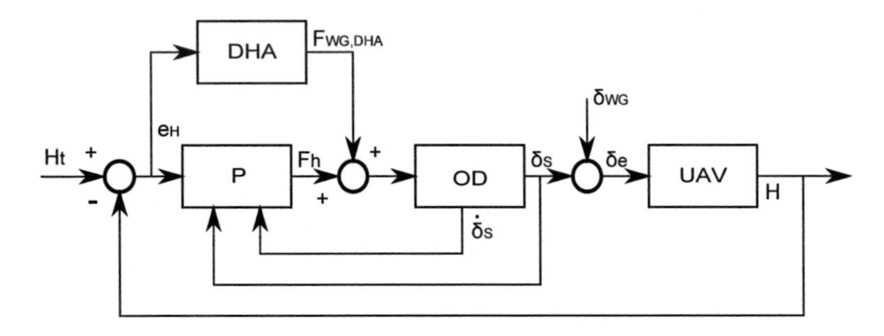

Figure 3.12. Compensator-Based DHA simulator scheme.

The *DHA* block in Figure 3.12 is a compensator represented by the transfer function of Equation (3.28) which calculates the DHA external force starting from the altitude error. It is designed in order to damp the Phugoid mode like a good pilot would do and cancel the Omega Device dynamics (see Section C.1). In order to design the DHA compensator, the Omega Device dynamics was identified first (see Section B for details). The net result is that such compensator can damp effectively the Phugoid mode from altitude measurement by itself, without any pilot in the loop: the stick moves and the corresponding stick deflection is sufficient to control the aircraft. In order to leave the pilot with enough control authority, the gain of the compensator was reduced by 60%:

$$\frac{F_{WG,DHA}(s)}{e_H(s)} = \frac{6452s^2 + 2584s}{s^4 + 14.75s^3 + 209.5s^2 + 1089s + 13.04} \tag{3.28}$$

Thus, the force felt in DHA case is given from Equation (3.20) by considering $F_{WG,fi} = F_{WG,DHA}$ of the Equation (3.28):

$$F_{DHA} = F_{SD,x} + F_{WG,DHA} \tag{3.29}$$

Note the Spring-Damper component is the same in each of the three force conditions (NoEF, IHA and DHA). Suppose a downward wind gust affects the aircraft: the altitude tends to decrease. Within this condition, the DHA compensator produces a positive force (see Figure 3.2), $F_{WG,DHA}$, that would produce a positive stick deflection, δ_S, and thus induces the aircraft to climb back to the target altitude (the initial one). In this specific case the pilot feels a force that pulls the stick, that is to climb, and he/she should be compliant with the force, by following and amplifying the stick motion, in order to keep the altitude constant. This type of force feedback, roughly speaking with the same sign than the actual maneuver to be taken, is in complete accordance with the DHA concept.

Figure 3.13 depicts an example of the variables history during a simulation trial.

The design of a DHA based augmentation scheme seems to be very task-dependent; the compensator-based design approach described above is viable in this case since the task is specified as holding a reference altitude. Such an approach could not be used instead when the task cannot be specified as a reference signal to be tracked, or the pilot intention is not known; thus the design of a DHA augmentation scheme

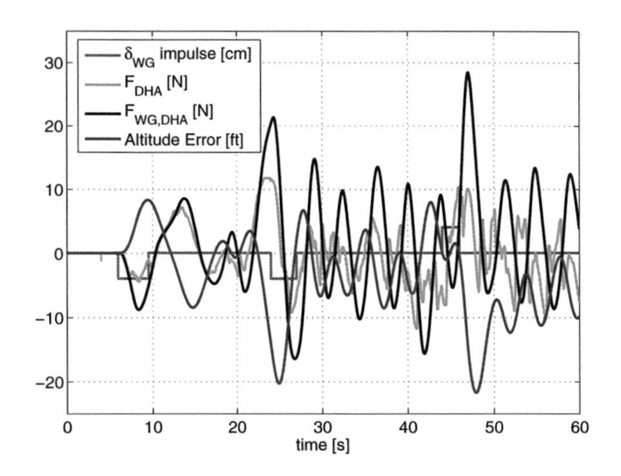

Figure 3.13. DHA simulation example.

could be less straightforward than an IHA one.

The Section 3.6 presents the experimental evaluation of the CAAF concepts.

3.6 CAAF Evaluation

This Section presents the experimental evaluation of the concepts described in Sections 3.3 and 3.5. In particular, the Sections 3.6.1 and 3.6.2 describe the **CAAF Experiment** and results and the Sections 3.6.3 and 3.6.4 describe the **CAAF VS DHA Experiment** and results.

3.6.1 CAAF Experiment

Within the CAAF Experiment, object of this section, a simple regulation task was prepared: the aircraft is initially flying leveled in trim conditions

($300 ft$ altitude) and at constant altitude; at a certain time, a disturbance (elevator impulse) is artificially injected and the aircraft initiates a motion according to its Phugoid mode.

The pilot's task is to keep the aircraft leveled, non oscillating, to restore the initial altitude and to keep it as constant as possible. During this task, the pitch and altitude oscillations of the Phugoid mode have to be damped by the pilot using the stick (as the red line in Figure 3.5).

The goal of these tests is to prove whether adding the Variable Stiffness CAAF kinesthetic (force) cue to the visual cue (a simulated cockpit) improves the control. In particular the goal is to assess as analytically as possible the differences in pilot performance in the two cases: with and without Variable Stiffness CAAF; the performance of the subjects (dependent variable) is measured through the IAE (Integral Absolute Error) between the current altitude and the desired one; a smaller IAE would indicate a better pilot performance in damping the Phugoid mode.

Eighteen naïve subjects (aged 23 to 43, mean 30.7) participated to the experiment. All had normal or corrected-to-normal vision. They were paid, naïve as to the purpose of the study, and gave their informed consent. The experiments were approved by the Ethics Committee of the University Clinic of Tübingen, and conformed with the 1964 Declaration of Helsinki. The experiment consisted of three different force conditions: No Force condition or *NoF*, with only compensation of gravity activated on the end-effector, Simple Force condition or *IHA-VS CAAF*, the Variable Stiffness CAAF of Equation (3.19), and the Double Force condition or *IHA-Double VS CAAF*, twice as much as the force in the Simple Force condition, achieved by doubling the $K_{f,vs}$ gain in Equation (3.19). Each condition was run as a separate block, i.e., the experiment consisted of three successive blocks. The order of presentation of the blocks was counterbalanced (see Section D.1 for details).

In total, the experiment lasted from 60 to 90 minutes (including instructions and breaks between blocks).

3.6.2 CAAF Experimental Results

Mean IAE values were entered in a one-way repeated measures analysis of variance (ANOVA) [NoF, IHA-VS CAAF, IHA-Double VS CAAF] (VS is for Variable Stiffness), which revealed a significant effect of the force factor

$$[F(2,34) = 7.932, p < 0.01]$$

As shown in Figure 3.14, the participants were the least variable (performed best) when a simple force was applied, the most variable (performed worst) when no force was applied, whereas providing a double force gave rise to 'intermediate' results.

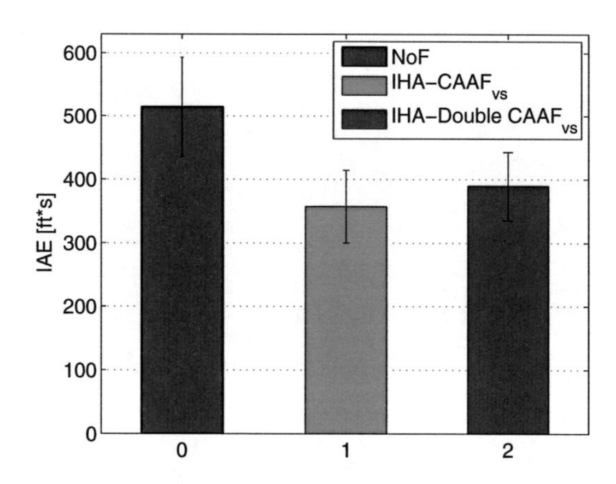

Figure 3.14. Performance (mean and standard error) for the three Force conditions (NoF, IHA-VS CAAF, IHA-Double VS CAAF).

Post-hoc tests using Bonferroni correction for multiple comparisons ($p < 0.05$) indicated that the performance with force (both Simple and

Double) was significantly less variable than without force. In other words, providing Variable Stiffness CAAF force significantly improved piloting performance as it reduced the variability of the control. It was also assessed the effect of the order of presentation of the blocks through a one-way repeated measures analysis of variance ANOVA [First Block, Second Block, Third Block], which revealed no significant main effect of the order of presentation. In other words, the variability of the performance was comparable irrespective of the order of presentation.

The results clearly show that the Variable Stiffness CAAF facilitates control in this task. Indeed, participants' performance significantly improved when haptic cue was available. As none of the participants had any experience with piloting, the results suggest that this type of aiding is rather 'natural', as beneficial effects can be observed without any previous learning. In line with these convincing initial results (published by the author in [2, 4]), could be interesting as future work to investigate the amount of additional information transferred to the operator via the CAAF variable stiffness haptic feedback as compared with other types of haptic aids (e.g., constant stiffness).

3.6.3 CAAF VS DHA Experiment

Within the CAAF VS DHA Experiment, object of this section, a simple control task was prepared: the aircraft is initially flying leveled in trimmed condition at constant altitude ($300ft$); three severe vertical wind gusts, which induce the aircraft to initiate a motion according to its Phugoid mode, are simulated by artificially injecting three control disturbances (elevator impulses) of randomized duration (2, 3 or 3.5 seconds), starting time and sign (upward or downward). See Section D.2 for details.

During this task, the pitch and altitude oscillations of the Phugoid mode had to be damped by the pilot through the use of the stick.

When a vertical wind gust disturbance affects a manned aircraft, the change in angle of attack and wing load are almost instantaneous. This has also an immediate effect on a mechanical-linkage based control stick. The altimeter on the GCS cockpit will though show the resulting change in altitude with a certain delay with respect to the actual disturbance time; as a matter of fact the aircraft dynamics from angle of attack to altitude has a low pass behavior and phase lag (in the simplest linear approximation it behaves as an integrator).

In order to focus on the haptic cues the experiment was made more difficult for the pilots by setting the Artificial Horizon inoperable (zero pitch and roll); only altitude and speed readings were displayed.

The experiment consisted of three different external force conditions: No External Force condition (referred as *NoEF* condition) with only the spring-damper force on the end-effector, IHA condition (the Force Injection CAAF from Equation (3.21)) and DHA condition (see the Equations (3.29) and (3.28)).

To summarize the forces felt by the pilots during the experiment an example is given: when a vertical wind gust (upwards for example) affects the aircraft, it will climb. The pilot should push over in order to reject the gust. So, to reject the gust the pilot should be compliant with the DHA Force and should oppose to the IHA Force.

All the trials were mixed and counter-balanced (see Section D.2 for details) and no instructions were given about the three different force conditions to test the natural reaction of the pilots to the three different conditions.

A test campaign with professional pilots was performed for the altitude regulation task. Seven professional pilots (from 50 to 700 hours of flight experience) participated to the experiment. The goal of these tests was to prove whether adding the IHA-Force Injection CAAF kinesthetic (force) cue or the DHA kinesthetic (force) to the visual cue (the simulated cockpit), improved the control with respect to a simple spring-damper behavior of the stick (NoEF). In particular the goal was to assess as analytically as possible the differences in pilot performance in the three cases identified as NoEF, IHA-Force Injection CAAF, DHA. The performance of the subjects (dependent variable) was measured through the IAE (Integral Absolute Error) between the current altitude and the desired one; a smaller IAE would indicate a better pilot performance in damping the Phugoid mode.

All the trials (36 of 60 seconds each, 12 trials per condition) were, as said, mixed and counter-balanced to test natural reaction of the pilots to the three different conditions. Before starting the experiment, every pilot was asked to run a 5 minutes trial where he/she had to perform a slightly different altitude regulation task; the goal of this initial trial, was to let the pilot acquire enough knowledge of aircraft dynamics to be able to pilot it confidently. During this trial a simple spring-damper (stiffness and damping constants were chosen as 1/6 of the NoEF case) behavior of the stick was employed. In total the experiment lasted 90 minutes. All pilots had normal or corrected-to-normal vision; they were paid and gave their informed consent. The experiments were approved by the Ethics Committee of the University Clinic of Tübingen, and conformed with the 1964 Declaration of Helsinki.

3.6.4 CAAF VS DHA Experimental Results

Mean IAE values for the three force conditions [NoEF, IHA, DHA] were entered in a one-way repeated measures analysis of variance (ANOVA). When all trials (12 trials for each condition) were considered, no main effect of the type of force was observed, i.e., the three types of force did not differ from one another. It was then assessed whether all three types of force feedback were equally 'natural' for the subjects, i.e., whether the first exposure to the different types of feedback gave rise to comparable performance. Here, only the first two trials of each subject for each condition were considered, and the data were entered in the same one-way ANOVA (described above). This analysis revealed a main effect of the type of force feedback

$$[F(2,12) = 12.943, p < 0.01]$$

As shown in Figure 3.15, the participants were the least variable in the NoEF and IHA conditions, and the most variable when the DHA force was applied, being the variability significantly worse in this last condition (post-hoc tests using Bonferroni correction for multiple comparisons, $p < 0.05$). In other words, when completely naïve about the aiding schemes (in the first two trials), participants performed significantly better when either no force or the IHA aiding scheme was used than with the DHA aiding scheme.

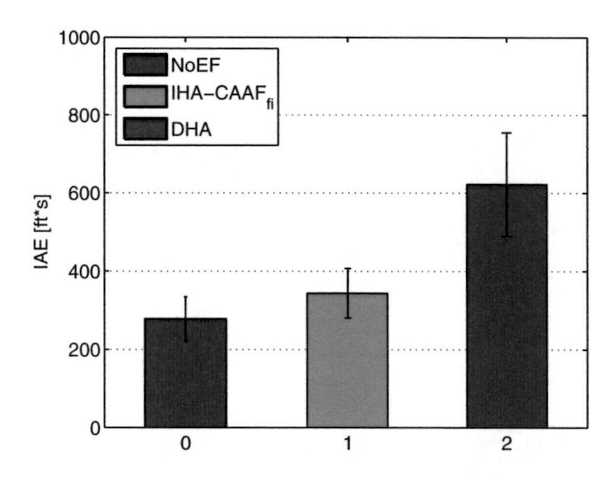

Figure 3.15. Performance (mean and standard error) for the 3 Force conditions of the first 2 trials.

Assuming that a certain degree of adaptation and learning of the pilots could have happened during the 12 trials, the last five trials of each condition were also evaluated separately. To test whether this was the case, the mean values of the last five trials were entered in the same one-way ANOVA. The analysis revealed a main effect of the type of force feedback

$$[F(2,12) = 13.007, p < 0.001]$$

As shown in Figure 3.16, the participants were the least variable when the DHA force was applied, and the most variable when both NoEF and IHA forces were applied. Post-hoc comparisons using Bonferroni correction ($p < 0.05$) showed that this difference was significant. In other words, after some training, the DHA approach allowed the best results. It is worth noticing that, the pilot were not trained explicitly on the three force conditions, and that the trials consisted of a sequence of mixed conditions and not of a uniform batch of the same force condition; thus, although no training was provided to the pilots on any of the three conditions, the pilots were quickly capable of understanding the DHA functionality and exploited it for improving their performance.

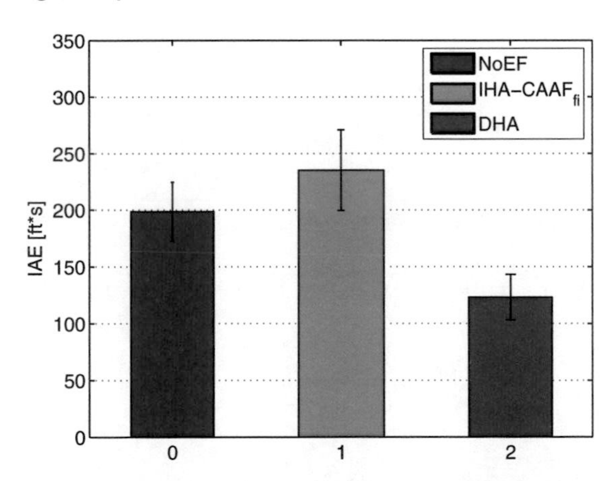

Figure 3.16. Performance (mean and standard error) for the 3 Force conditions of the last 5 trials.

After each experiment, pilots were interviewed separately; first of all the pilots were asked to describe their experience and identify the number of different types of sensations they felt during the experiment. All of them identified mainly two classes of force feedback: one which they called "natural", another one which they called "autopilot"; as a matter of fact, after few tests (from 2 to 4), they realized that in certain experiments the system was providing forces that where oriented in the direction of helping to perform the maneuver (autopilot case) and in other cases the forces were easier to associate with what they were expecting as the aircraft behavior (natural case). Only one pilot realized that some trials were run with the no force case in which the external disturbances gave no sensation through the stick. Thus, in order to compare the results, each pilot was asked to fill out a questionnaire with 6 questions (Table 3.1). In each question he/she had to choose, accordingly to the classification of sensations described above, between two different force feedback cases: "Natural" and "Autopilot". According to the discussions with the pilots, the author is confident that the Natural case can be mapped to the union of the NoEF and IHA cases, while the Autopilot case can be mapped to the DHA condition. The 6 questions in the questionnaire are shown in the Table 3.1:

A.	*Which force condition was stronger?*
B.	*Which of the two conditions do you think was more helpful?*
C.	*Under which condition you think you had the best control on the aircraft?*
D.	*In which condition you think you had to produce the largest effort?*
E.	*In which of the conditions you think you had the best performance?*
F.	*Which of the conditions did you prefer?*

Table 3.1. The wind gust rejection task questionnaire.

Figure 3.17 shows the corresponding pilot answers.

Most pilots agreed that the Autopilot case presented stronger forces and was more helpful (Questions A and B) than the Natural case. Answers

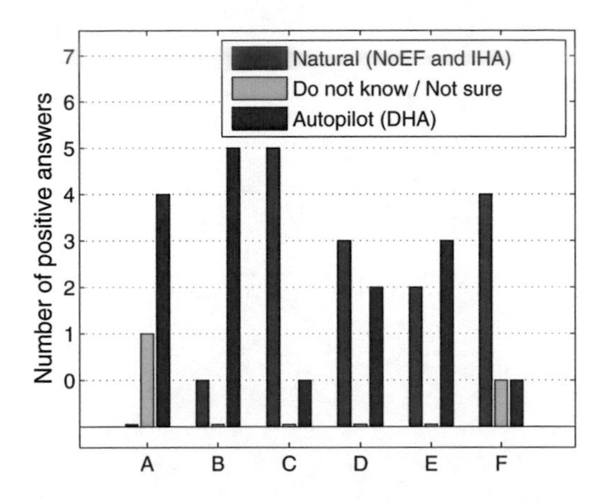

Figure 3.17. Pilot answers to questionnaire.

to question B and C showed a controversial situation: although most pilots voted for the Autopilot as the most helpful, most pilots felt more like being actually piloting the aircraft (Question C) with the Natural case. Pilots' opinions about the workload (Question D) and about the evaluation of their own performance in the task (Question E) were divided. Finally, although it could appear that pilots were going to prefer the Autopilot case, most of them voted for the Natural case. Regarding the latter question, the pilot who voted "not sure" said that he would have voted for the Autopilot case but after a longer training (the author published the results in [3, 4]).

By concluding, the NoEF and IHA conditions are the most natural forces to the pilots while after some training they can adapt to the DHA force feedback producing the best results even if the workload in this case results to be greater than in the previous ones.

4 Obstacle Avoidance Feel

According to [53], the haptic feedback can compensate to some extent for the lack of sensory cues that will be presented to UAV operators (see Section 1.3); this means that the addition of a haptic interface to the usually employed visual interface may improve the situational awareness of a remote UAV pilot and the efficiency of the teleoperation. It seems to be particularly necessary in cases of limited visual information. In the presence of foggy weather conditions, for example, or because of the employment of a limited FOV camera [40], the haptic feedback could provide information through the sense of touch which can be applied directly on the control device.

In Section 2.2 a classification about the haptic aids present in scientific literature is given. The haptic aids, as said, are classified in DHA and IHA. Most of haptic literature is based on DHA concept. In particular, as concerning the obstacle avoidance task every existing haptic aid seems to belong to the DHA class. Usually in this class a repulsive force is associated to the obstacles. Thus the pilot (or the human operator in general) has to be compliant with the force felt on the control device. In this Chapter an attempt of designing a force feedback belonging to the IHA class for the obstacle avoidance task is made. This will be shown to be more "natural" than the usually employed DHA-based approaches confirming what Schmidt and Lee [73] expatiated (see Section 2.2).

The research resulted in an obstacle avoidance/detection force named *IHA-Obstacle Avoidance Feel* (IHA-OAF) and it is the object of the present Chapter (the author published the results in [1, 5]). It will be shown to definitely improve the pilots' sensations and performance!

4.1 Simulation Environment

The present Section describes in details the simulation environment of the obstacle avoidance task.

Figure 4.1 shows the setup employed in the experiment. The virtual environment display produces the visual cue: a subjective view from the aircraft cockpit is simulated using a realistic virtual environment created through the DynaWORLDS software package [58] (see Section A.3 for details on the implementation). The environment is constituted by a non-Manhattan scenario (see Figure D.3) with a ground plane, the sky and buildings with regularly spaced windows to reproduce an appropriate perception of depth. As a matter of fact, the teleoperation of a vehicle in a opened area makes the simulation less problematic than the implementation of a constrained environment as long as, in the latter case, an accidental reduction of the visual feedback or small delays could bring to collisions. The obstacle avoidance task is a challenging problem in robotics.

To make the implementation of the experiment easier, the full non linear dynamics previously mentioned (DHC-2 Beaver [68]) is linearized around the trim conditions (horizontal flight at $300 ft$ altitude). As concerning

Figure 4.1. The obstacle avoidance teleoperation setup.

the obstacle avoidance task the aircraft dynamics is decoupled and only the lateral dynamic is considered. The Equation (3.5) shows how the elevator deflection through the changes of the lift coefficients modifies the aircraft lift and thus the longitudinal aircraft trajectory. It concerned the longitudinal dynamics. As concerning the lateral dynamics, something similar to Equation (3.5) can be written.

In order to limit pilot workload and possible errors, only the aircraft lateral dynamics (i.e. roll and heading angles and lateral position) has to be controlled by the pilot. Equation (4.1) shows the lateral steady state equations of horizontal flight in Wind Axes (see Section 3.2.2 for the definition) [78]:

$$\begin{cases} 0 = C = C_C \cdot qS \\ 0 = l = C_l \cdot bqS \\ 0 = n = C_n \cdot bqS \end{cases} \tag{4.1}$$

where C, l and n are respectively the aircraft cross-wind force, the rolling and the yawing moments; C_C, C_l and C_n are respectively the aircraft cross-wind force, rolling and yawing moment coefficients. q, b and S are the dynamic pressure, the wing span and surface respectively. As in Equation (3.5), the coefficients present in Equation (4.1) can be re-written as proportional to the sideslip angle β, the angle between the aircraft direction of the motion (the relative speed) and the xz-plane in the Body Reference Frame (see Section 3.2.1 and Section 3.2.2 for the definition), and to the aileron deflection δ_a (the rudder deflection δ_r and α are supposed to be fixed in the respective trim condition values). The aileron deflection δ_a is supposed to be the only input to the lateral dynamics. It is proportional to the lateral stick deflection δ_A (in this work $\delta_a = \delta_A$) and can be thought as generated by the operator by moving the haptic device end-effector in the y-direction (see Figure 3.2). An input on the aileron produces the aircraft lateral motion. See the following Section for details on the lateral linear dynamic model of the aircraft employed.

4.2 Aircraft Lateral Dynamics

Figure 4.2 shows the baseline scheme (i.e. no haptic aids) employed in the obstacle avoidance setup.

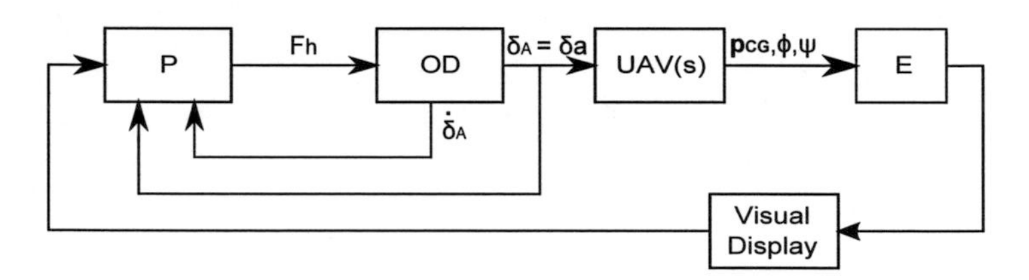

Figure 4.2. The obstacle avoidance simulation baseline scheme.

The input of the aircraft lateral dynamics $UAV(s)$ is the aileron deflection, δ_a, (in this Chapter coincident with the lateral deflection δ_A of the haptic device represented by the OD block) and the outputs are the aircraft center of gravity position $\mathbf{p}_{\mathbf{CG}}$ or (x_e, y_e), heading (ψ) and roll angle (ϕ) of the aircraft in the Earth Reference Frame (Earth-Centered, Earth-Fixed reference frame with origin in the center of the Earth, z_{OB} axis points North, x_{OB} and y_{OB} axes are on the equatorial plane). The block $UAV(s)$ of Figure 4.2 is shown in details in Figure 4.3.

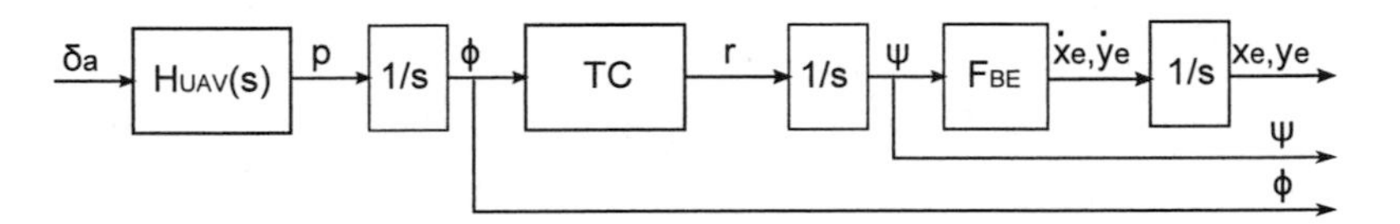

Figure 4.3. The aircraft lateral dynamics.

In Figure 4.3 the transfer function $H_{UAV}(s)$ from aileron, δ_a, to roll rate, p or $\dot{\phi}$ (Equation (4.2)) is employed. It is obtained through linearization of the non linear Beaver DHC-2 of the *Flight Dynamics and Control Toolbox* [68]. The roll angle, ϕ, is obtained via integration and saturated to $50 degrees$ to make the aircraft dynamics more realistic.

$$H_{Lat} = \frac{-9.9877s^3 - 10.4132s^2 - 6.1385s + 0.0184}{s^4 + 8.1578s^3 + 10.2490s^2 + 11.8186s + 0.6961} \tag{4.2}$$

Then, by making the assumption of aircraft performing a coordinated turn [69] (TC block in Figure 4.3) (zero velocity in the lateral body axis) at constant speed V (about $50m/s$), the heading rate r or $\dot{\psi}$ is calculated through the Equation (4.3):

$$\dot{\psi} = r = tan(\phi)\frac{g}{V} \tag{4.3}$$

where g is the gravity acceleration. The heading angle, ψ, is obtained via integration. $\dot{x_e}$ and $\dot{y_e}$ are calculated through coordinates transformation (F_{BE} block) of Equation (4.4) from Body Reference Frame to Earth Reference Frame (see the Section 3.2.1 and above for the reference frames definitions).

$$\begin{bmatrix} \dot{x}_e \\ \dot{y}_e \end{bmatrix} = \begin{bmatrix} cos(\psi) \\ sin(\psi) \end{bmatrix} V \tag{4.4}$$

The coordinates of the aircraft center of gravity (x_e and y_e) are calculated from Equation (4.4) via integration. As shown in Figure 4.2, informations about the environmental constrains (contained in E block) are used to show through the *Visual Display* (see Figure 4.1) the virtual environment from the camera point of view which is represented by the aircraft center of gravity position and attitude.

4.3 The Stick Force

Since, by hypothesis, the only dynamics to control is the lateral one, the haptic aid for the obstacle avoidance task is applied only to the control device lateral axis (the y axis in Figure 3.2).

Thus, the only force transmitted to the operator is along the y axis.

$$\begin{cases} F_y = F_{SD,y} + F_{OA,y} \\ F_{SD,y} = F_{SD} = F_{S,y} + F_{D,y} = K_{S,y} \cdot y_S + K_{D,y} \cdot \dot{y}_S \\ F_{OA,y} = F_{OA} \end{cases} \tag{4.5}$$

In Equation (4.5), F_{SD} is the Spring-Damper force. Lateral stiffness and damping coefficients, $K_{S,y}$ and $K_{D,y}$, are chosen as in Table A.2. The lateral stiffness is one half of the longitudinal stiffness. As a matter of fact, as concerning the force/displacement characteristic, the sticks have usually stiffer gradients pitching commands (forward/backward arm movement) than rolling commands (left/right arm movements) [82] because of the differences in strength among the various arm muscles used for pitch and roll control. Similar difference exists between pulling movements (both longitudinal and lateral) towards the pilot body and pushing movements away from it [82] but in this work the stiffness is supposed to be constant for both longitudinal and lateral movements although the different values (smaller for lateral stick displacements).

The Spring-Damper term depends on the desired stick dynamics and it is present (same value) in all the conditions of the experiment, while the external force term for the obstacle avoidance, F_{OA}, depends on the experimental conditions: three types of external force F_{OA} are compared: DHA, IHA and a baseline force condition (No External Force, NoEF) in which $F_{OA} = 0$ in order to test the operator's performance in the obstacle avoidance task. To create Direct and Indirect external forces two simulators are prepared (see Section 4.4).

4.3.1 The haptic feedback

It is well known that an aircraft (even the modern fly-by-wire ones) stick should always offer a certain stiffness and damping to the pilot to mimic a mechanically driven aircraft stick [69, 34]. In most teleoperation situations, it is common trying to make the haptic interface invisible to the human operator for achieving what is often defined as *transparency* of the teleoperation system. In this specific case though, it is believed that the user must always feel a certain interface stiffness and damping even when not *feeling the presence of the environment*. The author also proved the importance of the spring-damper force (as shown in Chapter 3) in a previous paper [2].

Thus, for this particular application, it is designed a system where the haptic interface appears as a stick with constant damping and stiffness with the addition of an external force when needed (namely when close to obstacles). Then, the force F_y felt by the operator during the obstacle avoidance task (see Equation (4.5) and (4.6)) along the control device y axis (see Figure 3.2) is a combination of an elastic term $F_{S,y}$, $(K_{S,y} \cdot y_S)$ with constant stiffness $K_{S,y}$, a damping term $F_{D,y}$, $(K_{D,y} \cdot \dot{y_S})$ with a damping constant $K_{D,y}$, and an external force component F_{OA}. y_S and $\dot{y_S}$, in the following indicated as δ_A and $\dot{\delta}_A$, are the lateral displacement and displacement rate of the end-effector respectively.

$$F_y = K_{S,y} \cdot y_S + K_{D,y} \cdot \dot{y_S} + F_{OA} \tag{4.6}$$

As said, the external force F_{OA} could belong either to the DHA class or to the IHA class. In the baseline force condition (No External Force, NoEF) $F_{OA} = 0$. Section 4.4 describes DHA and IHA obstacle avoidance external forces.

4.3.2 The Obstacle Force Field

In order to produce some kind of haptic feedback on the stick with the goal of helping to avoid collisions with obstacles, a force field around the obstacles (Equation (4.7)) is defined. The force field starts in the center of each single obstacle and points away from the obstacles.

The intensity of the force field decreases with distance from the obstacle and becomes zero beyond a certain threshold distance. A haptic sensation is thus produced as proportional to this force field.

The total force $\mathbf{F_{OBS}}$ exerted by the environment at the position of aircraft center of gravity, in the obstacle reference frame (Equation (4.7)), the fixed Earth Reference Frame (see Section 4.2 for the definition), is the superposition of the repulsive forces produced by each obstacle.

$$\mathbf{F_{OBS}} = \begin{bmatrix} F_{OBS,x} \\ F_{OBS,y} \end{bmatrix} = \sum_{n=1}^{N} \mathbf{F_{OB}} \tag{4.7}$$

where N is the total number of obstacles. For both DHA and IHA approaches, the force field shows a maximum intensity on the obstacle boundary decreasing with distance from it. The force field is present inside the obstacle as well (see later).

By following this principle, a repulsive force field (Equation (4.8)), similar to the one chosen by Melchiorri [15] and which represents the repulsive force field often used in literature, is associated to a collection of rectangular obstacles.

$$\mathbf{F_{OB}} = \begin{cases} -k_f \cdot (d(\mathbf{p_{OB}}, \mathbf{p_{CG}}) - r_e) \cdot \frac{\mathbf{p_{OB,C}} - \mathbf{p_{CG}}}{\|\mathbf{p_{OB,C}} - \mathbf{p_{CG}}\|}, & \\ & \textit{for } d(\mathbf{p_{OB}}, \mathbf{p_{CG}}) < r_e \\ 0, & \textit{otherwise} \end{cases} \tag{4.8}$$

Let $\mathbf{p_{CG}}$, $\mathbf{p_{OB,C}}$ and $\mathbf{p_{OB}}$ be respectively the position of the aircraft center of gravity (x_e, y_e), the position of the center of a single obstacle and the sides of the obstacle closer to the aircraft. In particular, the distance $d(\mathbf{p_{CG}} - \mathbf{p_{OB}})$ between the aircraft center of gravity and the obstacle depends on the position of the aircraft center of gravity with respect to the obstacle (see Figure 4.4). In particular, the aircraft can be positioned (see Figure 4.4) next to the obstacle sides (either *A* or *B* zone) or next to the obstacle vertices (*C* zone).

Depending on this, the distance between the aircraft center of gravity and the obstacle is defined as:

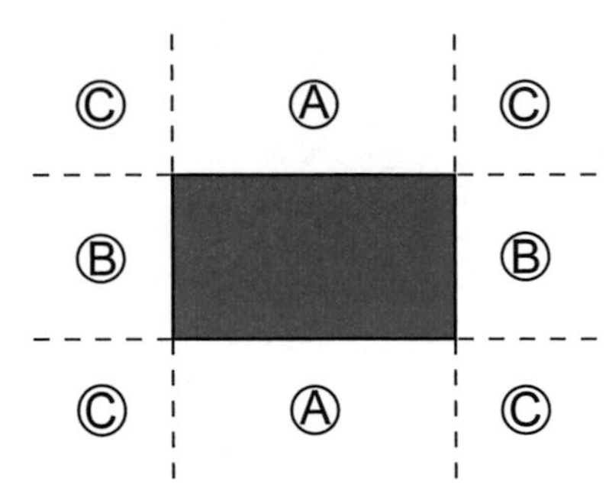

Figure 4.4. Definition of the distance between the aircraft center of gravity and the obstacle.

- **CASE A**: the vertical distance between $\mathbf{p}_{CG}$ and the closer horizontal obstacle side;
- **CASE B**: the horizontal distance between $\mathbf{p}_{CG}$ and the closer vertical obstacle side;
- **CASE C**: the euclidean sum of the previous ones;

The term $\frac{\mathbf{p}_{OB,C}-\mathbf{p}_{CG}}{||\mathbf{p}_{OB,C}-\mathbf{p}_{CG}||}$ indicates the force field versor given by the congiunction between the aircraft center of gravity and the center of the obstacle.

The force field at the position $\mathbf{p}_{CG}$ is aligned with the versor $\frac{\mathbf{p}_{OB,C}-\mathbf{p}_{CG}}{||\mathbf{p}_{OB,C}-\mathbf{p}_{CG}||}$ and the intensity is selected to be linearly decreasing with the distance $d(\mathbf{p}_{OB},\mathbf{p}_{CG})$ of the point $\mathbf{p}_{CG}$ from the nearest point of the obstacle boundary.

The constant k_f is an appropriately selected constant and can be thought as the stiffness of the virtual environment. When the distance $d(\mathbf{p}_{OB},\mathbf{p}_{CG})$ is less than r_e (set to $50m$), the maximum distance of influence, a repulsive force is used to generate the Haptic Aid in order to help the aircraft pilot in avoiding the obstacle.

Figure 4.5 shows an example of the force field with force vectors and ISO-force contour lines that is produced by the obstacles. The value and

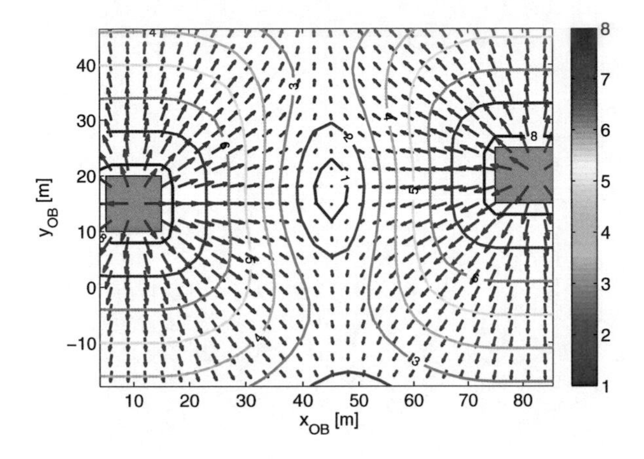

Figure 4.5. Example of the obstacle repulsive force field with contour lines. Note the low amplitude force field in the middle of the obstacles.

direction of the force field at the current position of the aircraft are used in the simulator to generate the haptic perception.

An example of the mentioned non-Manhattan scenario generated force field with force vectors and ISO-force contour lines is depicted in Figure 4.6.

Figure 4.6 clearly shows a low amplitude force field in the virtual corridor created in the middle of the street and the maximum force (about $10N$) at the obstacles sides.

As mentioned, the total force exerted by the obstacles (Equation (4.7)) is expressed in the fixed Earth Reference Frame. A change in the aircraft Body Reference Frame (see Section 4.2 for the definition) is necessary to appropriately select the force component that lies on the lateral axis of the

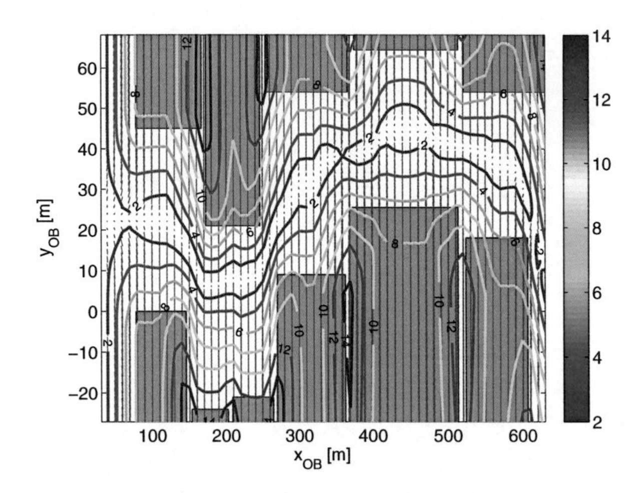

Figure 4.6. Example of non-Manhattan scenario repulsive force field with contour lines. Note the low amplitude force virtual corridor in the middle of the street.

current aircraft direction:

$$\begin{bmatrix} F_{B,x} \\ F_{B,y} \end{bmatrix} = \begin{bmatrix} cos(\psi) & sin(\psi) \\ -sin(\psi) & cos(\psi) \end{bmatrix} \cdot \begin{bmatrix} F_{OBS,x} \\ F_{OBS,y} \end{bmatrix} \tag{4.9}$$

In Equation (4.9), $F_{B,x}$ and $F_{B,y}$ are the force components in the aircraft Body Reference Frame. ψ is the aircraft heading angle. From now on, the force produced by the environment, F_{AO} of Equation (4.5), will be considered to coincide with the y component in Equation (4.9), i.e. $F_{OA} = F_{B,y}$.

4.4 The OAF VS DHA Experiment Simulators

In order to test the IHA-Obstacle Avoidance concept, three simulators are set up:

- the NoEF Simulator (see Section 4.4.1);
- the DHA Simulator (see Section 4.4.2);
- the IHA-OAF Simulator (see Section 4.4.3).

As preliminary assessment of the techniques and for tuning the IHA and DHA simulators, a simple experiment with an isolated obstacle is run (Section 4.4.4). A more complex scenario (the mentioned non-Manhattan scenario) is used instead for a deep test campaign (see Section 4.5).

4.4.1 NoEF Simulator

Figure 4.2 shows the block diagram of the simulation system used to test the NoEF feedback force that is the baseline scheme (no haptic cues related to the obstacles).

Let $\mathbf{p_{OBS}}$ represent the position of the obstacles and $\mathbf{p_{CG}}$ the position of the aircraft center of gravity. The pilot may perceive the distance from the obstacles using the visual display (see Figure 4.1). The pilot force input (F_h), is fed to the haptic device (OD block in Figure 4.2) to produce the stick deflection δ_A (which in this Chapter is used directly as aircraft aileron control δ_a by hypothesis). δ_A and $\dot{\delta}_A$ feedback indicate the proprioceptive feedback.

This case represents just a visual aid being the haptic feedback only related to the actual stick displacement and to its rate and not related instead to the environmental constraints.

The NoEF condition presents a constant stiffness stick (to simulate a fly-by-wire like situation). In the NoEF condition the force exerted by the haptic device is the same as in the Equation (4.6) except for F_{OA} which is set to zero in this condition. The pilot has the mentioned virtual scenario as the only instrument showing the virtual buildings from the aircraft center of gravity point of view (see Figure 4.1). The visual feedback is the same in all the conditions of the experiment.

Thus, the force felt in NoEF case is given from Equations (4.10) and (4.5) with $F_{OA} = 0$:

$$F_{NoEF} = F_{SD,y} \tag{4.10}$$

Suppose the aircraft is close to an obstacle: as long as $F_{OA} = 0$ in the Equation (4.5), no force is directly linked to the the obstacle. Thus, the pilot will not feel any haptic information about the presence of the obstacle but he/she will just see it through the visual display (only in good visibility conditions, i.e. no foggy weather) while approaching. The only haptic feedback felt by the pilot is proportional to δ_A and $\dot{\delta}_A$ produced only by the own input force F_h.

Figure 4.7 depicts an example of the variables' history during a simulation trial.

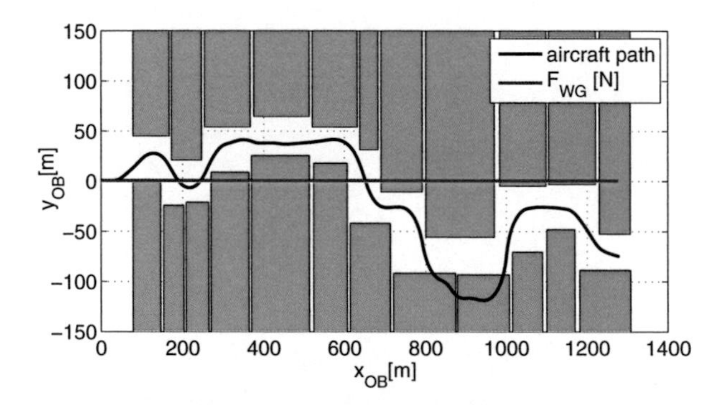

Figure 4.7. NoEF simulation example.

4.4.2 DHA Simulator

According to the DHA concept, a Direct Haptic Aiding system for obstacle avoidance should produce a force or a change in stiffness that "directly" helps the pilot in achieving the task that, in this case, is to avoid collisions with the obstacles. Thus, a system that produces a force which pulls the stick in the same direction the pilot should pull to avoid the collision seems appropriate for a DHA control. As a matter of fact, the obstacle avoidance system described in [40, 43] works exactly according to this principle: stiffness variation, together with force feedback were investigated and found to be able to provide better results than single stiffness or force feedback [43]. Nevertheless, for the purposes of the present comparison, force feedback only is investigated and compared in this work.
A compensator (DHA block in Figure 4.8) is added to compute the external force to be felt by the pilot. The Haptic device is controlled as in Equation (4.6) to behave as a spring-damper system with an additional force F_{OA} from the y component of Equation (4.9) (remember in fact that $F_{OA} = F_{B,y}$).

Figure 4.8 shows the block diagram of the simulation system used to test the DHA concept.

The pilot may perceive the distance from the obstacles using the visual display (see Figure 4.1). The same distance is also perceived via a haptic display through the DHA block that implements the DHA force and feeds-back the force (F_{WG}) which, together with the pilot force input (F_h), is fed to the haptic device (OD block in Figure 4.8) to produce the stick deflection δ_A. δ_A and $\dot{\delta}_A$ feedback indicate the proprioceptive feedback.

Thus, the force felt in DHA case is given from Equations (4.11) and (4.5) by considering $F_{OA,DHA} = F_{B,y}$ of the Equation (4.9):

$$F_{DHA} = F_{SD,y} + F_{OA,DHA} \tag{4.11}$$

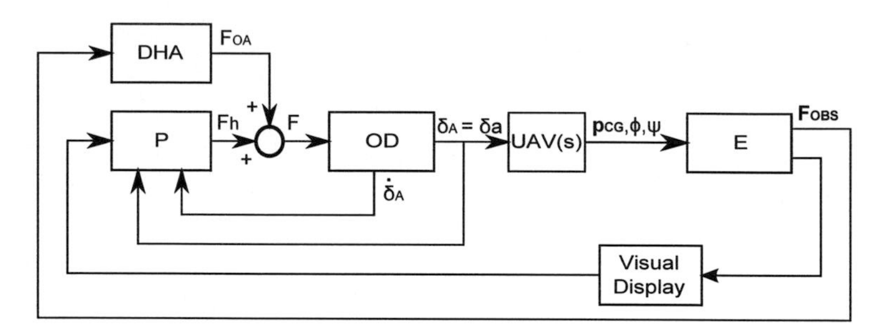

Figure 4.8. DHA-based obstacle avoidance simulator scheme. The haptic force F_{OA} deflects the stick inducing a helpful change of the aircraft trajectory.

Suppose the aircraft is close to an obstacle: being $F_{OA} = F_{B,y}$ in the Equations (4.5) and (4.6), the repulsive force F_{OA} generates a stick motion that deviates, at least partially, the aircraft trajectory away from the obstacle, thus the pilot has to follow it (being compliant) in order to avoid the collisions. The pilot will feel then a haptic information about the presence of the obstacle and he/she will see it through the visual display (when the visibility conditions are good enough) while approaching. A haptic feedback proportional to δ_A and $\dot{\delta}_A$ produced from both the pilot input force F_h and from the obstacle force F_{OA} is present as well.

Figure 4.9 depicts an example of the variables' history during a simulation trial.

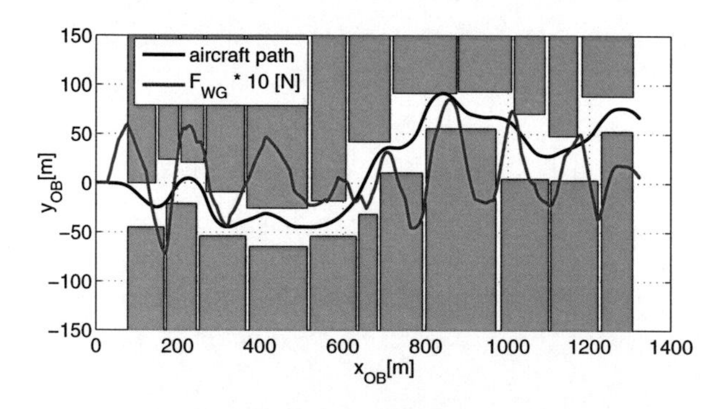

Figure 4.9. DHA simulation example.

4.4.3 IHA-OAF Simulator

The design of a IHA-inspired obstacle avoidance aid appears complex since no force sensation is "naturally" generated by coming close to an obstacle. Then, in order to follow the concept (already proved in Chapter 3 to be successful in the gust rejection task) that opposition to haptic stimuli is a "more natural" pilot reaction than compliance to stick motion (see Section 2.2), a haptic aid of opposite sign with respect to the DHA one was designed. This type of aid would result in a tendency of the aircraft to fly toward the obstacle instead of flying away from it as in DHA. Thus, for not penalizing too much the IHA system and to make it enough safe, the indirect force feedback (the same as the direct force feedback of Equations from (4.7) to (4.9) but opposite in sign) is transformed in a shift of the stick neutral point.

This means that only the stick, de facto, would move towards the obstacle without producing the aircraft to fly against it. For example, if an obstacle is on the right side, the stick would move to the right but, if the pilot is not in the loop, the UAV will continue to fly straight. What happens if the pilot is in the loop? In the same direction of what Schmidt and Lee think [73], the idea is that when the stick moves on one direction it would be more natural for the pilot to move it in the opposite side. Going back to the example: with the obstacle on the right, the neutral point of the stick shifts to the right, the pilot would feel this movement and perhaps he/she naturally would oppose it by moving the stick toward the left (that is simply moving the stick a little back to the center) performing a turn on the left that is, in the example, the maneuver to perform to fly away from the obstacle.

The vanishing of the haptic cue informs the pilot that the obstacle is far away and not dangerous anymore.

Figure 4.10 shows the block diagram of the simulation system used to test the IHA concept.

The distance between the obstacles and aircraft center of gravity may be perceived by the pilot P via the visual display (see Figure 4.1). The same distance is also perceived via haptic display through the IHA block that implements the IHA force and feeds-back the force F_{OA} which, together with the pilot force input (F_h), is fed to the haptic device (OD block in Figure 4.10) to produce the stick deflection δ_A. The block OD_i takes care of producing the effect of shifting the neutral point of the stick and will be detailed later. δ_A and $\dot{\delta}_A$ feedback indicate the proprioceptive feedback.

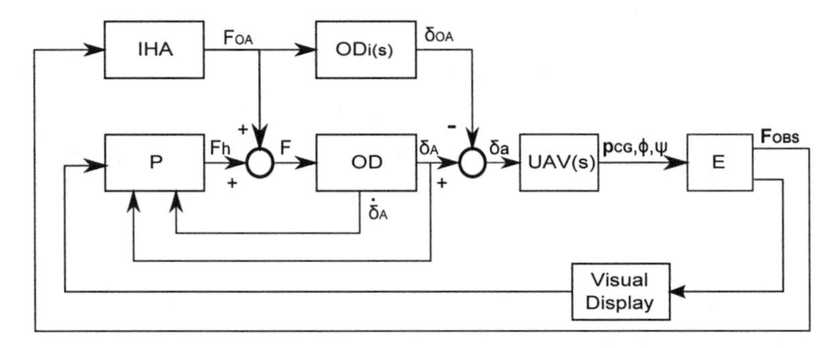

Figure 4.10. IHA-OAF simulator scheme. The haptic force F_{OA} deflects the stick without producing any change to the aircraft trajectory thanks to the effect of the compensating signal δ_{OA}.

Thus, the force felt in IHA case is given from Equation (4.12) by considering $F_{OA,IHA} = -F_{B,y}$ of the Equation (4.9):

$$F_{IHA} = F_{SD,y} + F_{OA,IHA} \tag{4.12}$$

Suppose the aircraft is close to an obstacle: being in this case $F_{OA} = -F_{B,y}$ of the Equations (4.5) and (4.6), a force which attracts the stick neutral point is directly linked to the the obstacles and the pilot has to oppose it in order to avoid the collisions. In fact, the shifting of the stick (neutral point) towards the obstacle makes the pilot to think that he/she is flying against the obstacle. The force is immediately felt by the pilot who learns that something changed. The pilot should react immediately by opposing to the stick motion in order to fly away from a possible collision.

Thus, the pilot will feel a haptic information about the presence of the obstacle and he/she will see it through the visual display (when the visibility conditions are good enough, i.e. no foggy weather) while approaching. A haptic feedback proportional to δ_A and $\dot{\delta}_A$, produced from both the pilot input force F_h and from the obstacle force F_{OA}, is present as well.

Figure 4.11 depicts an example of the variables' history during a simulation trial.

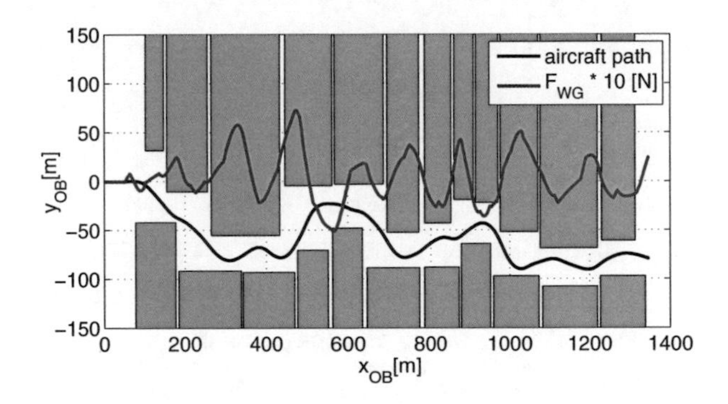

Figure 4.11. IHA-Obstacle Avoidance Feel simulation example.

In other words, the IHA-OAF follows the general IHA concept described before: it provides the pilot with the information about the presence of the obstacle on a side of the aircraft but it does not effect in any way the commands actually sent to the aircraft; this helps the pilot indirectly by improving his/her SA, that is to let him/her know that in the remote environment a collision is going to happen, and leaving him/her the full authority to take control decisions by changing the direction of the motion of the vehicle.

A mathematical proof of the neutral point shift concept described above is presented in the following Subsection.

IHA-OAF Implementation Proof

In order to modify the neutral point so that the haptic force F_{OA} would produce no actual change of the aircraft trajectory (i.e. the aircraft continues to fly straight if the pilot takes no command actions), the same external force, F_{OA}, is sent to both the real Haptic Device (actually the Omega 3DOF Device, Force Dimension, Switzerland) and an identified mathematical model of it (see Appendix B). The output of the identified haptic device model is subtracted from the total displacement of the end-effector of the real device in a way that the effect of F_{OA} will not be an input command to the aircraft but just a change in the neutral position of the stick.

Let $OD(s)$ be the transfer function of the real Omega Device (by supposing that the real Omega Device has a linear behavior and representing it through a transfer function is possible) and with $OD_i(s)$ the transfer function of the identified model of it. Let the displacement of the real Omega Device end-effector and the displacement of the identified model of it be respectively δ_{OA} and $\delta_{OA,i}$. Suppose that by giving the same input, F_{OA}, to the Omega Device and to its identified model the output, the produced displacement, is the same in both cases: $\delta_{OA} = \delta_{OA,i}$ (i.e. the identified model is exact); the net result is that the operator moves the end-effector by δ_A through the application of the force F_h. As a matter of fact, from the Figure 4.10:

$$F_h + F_{OA} = F \tag{4.13}$$

$$\delta_A - \delta_{OA} = \delta_a \tag{4.14}$$

$$\delta_{OA} = OD(s) \cdot F_{OA} = OD_i(s) \cdot F_{OA} = \delta_{OA,i} \tag{4.15}$$

$$\delta_A = OD(s) \cdot F = OD(s) \cdot (F_h + F_{OA}) = OD(s) \cdot F_h + \delta_{OA} \tag{4.16}$$

From the Equation (4.14) and the Equation (4.16):

$$\delta_A = OD(s) \cdot F_h \tag{4.17}$$

The final result is that the F_{OA} changes just the neutral point of the Omega Device of δ_{OA} and the only input to the aircraft dynamics is given by the pilot command F_h (Equation (4.17)). The transfer function $OD_i(s)$ of the actual Haptic device used in the experiments was identified by using frequency sweeps (from 0.0262 to 10 Hz) and the Empirical Transfer Function Estimate (ETFE) technique (Ljung, 1999) (see Appendix B).

4.4.4 Isolated Obstacle Scenario

In order to test the beneficial anticipatory effect of the haptic feedback and for tuning IHA and DHA simulators several experiments were run using a scenario with an isolated obstacle placed along the aircraft path; the task of the participant was to fly straight. The participant sees the obstacle from different distances, according to three visibility conditions (see Figure 4.13).

The most relevant test was run under the lowest visibility condition (i.e. foggy weather condition): the participant was not able to detect the presence of the obstacle early enough to maneuver the aircraft without the haptic feedback. As can be noted in Figure 4.12, while in the DHA and the IHA cases no collisions occurred, in the NoEF case a collision occurred confirming, at least according to this preliminary results, the importance of having a haptic feedback in addition to visual feedback to improve the flight safety. The increased reaction delay in the NoEF case, with respect to DHA and IHA, appears clearly from the F_y stick forces plots (magenta lines). In fact, the magenta line in NoEF case is null till the last $10m$ before the obstacle.

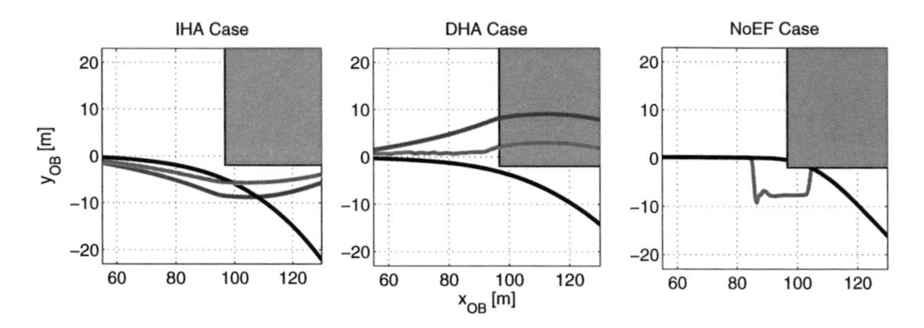

Figure 4.12. Isolated obstacle scenario: IHA, DHA and NoEF experiments in the Maximum Fog visibility condition. The lines represent: the aircraft trajectory (black) starting from the left, the force F_{OA} (blue when present) and the total force F_y (magenta).

4.5 IHA-OAF Evaluation

In order to test the IHA-Obstacle Avoidance concept, a deep test campaign on obstacle avoidance task in an urban-canyon complex scenario was run under three different visibility conditions: a) *Minimum Fog*; b) *Medium Fog*; c) *Maximum Fog* (see Figure 4.13) and under three different force condition: DHA, IHA and NoEF.

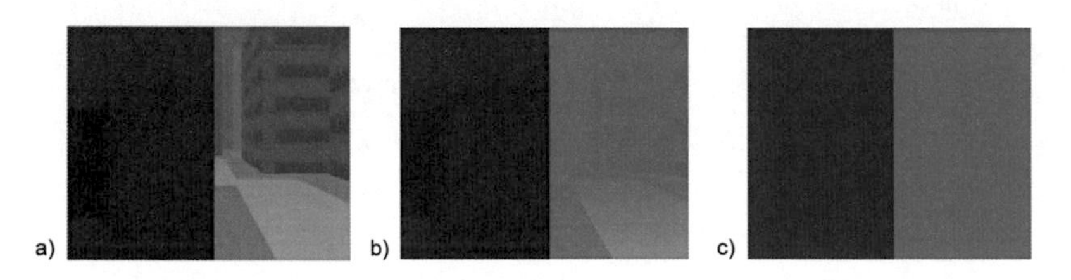

Figure 4.13. Out of the window view from the same viewpoint while the same obstacle, in the left side, is approaching under the three different visibility conditions: a) *Minimum Fog*; b) *Medium Fog*; c) *Maximum Fog*.

In Figure 4.13c the fog is so thick that the only information the pilot can rely on is the haptic cue only. Under the three different visual conditions, in fact, when an obstacle placed along the path of the aircraft, the pilot sees it from different distances and the available time to react for avoiding the collision is different. The most relevant test was run under the Maximum Fog visibility condition; in this case the pilot was not able to detect the presence of the obstacle early enough to maneuver the aircraft without the haptic feedback.

A simple control task was prepared: the aircraft had to be flown in an urban canyon with buildings placed irregularly (non Manhattan-like) along the desired path; thus, the buildings constituted a narrow street with buildings in both sides. The task of the experiment was to get the end of the street avoiding the collisions with them. Five different scenarios (i.e. position of the N obstacles) were used to avoid the effect of learning in test subjects (see Figure D.3 for an example about one of the 5 employed scenarios). To test the natural response to the different types of force no instructions were given to the participants about the force they were going to feel on the stick.

The error metric is the number of collisions.

The goal of these tests is to prove whether adding the IHA-OAF kinesthetic (force) cue to the visual cue (a simulated cockpit) improves the control with respect to the other two conditions. In particular the goal is to assess as analytically as possible the differences in pilot performance in the three cases. Thus, the performance of the subjects (dependent variable) was measured through the number of collisions in the flight across a constrained environment.

Ten naïve subjects participated to the experiment. All had normal or corrected-to-normal vision. They were paid, naïve as to the purpose of the study, and gave their informed consent. The experiments were approved by the Ethics Committee of the University Clinic of Tübingen, and conformed with the 1964 Declaration of Helsinki.

The experiment consisted of three different force conditions: NoEF, DHA and IHA-OAF.

All the trials (see Section D.3 for details) have been mixed and counter-balanced and no instructions were given about the three different force conditions to test natural reaction of the subjects to the three different conditions.

Each fog condition was run as a separate block, i.e., the experiment consisted of three successive blocks.

The participants in the experiment had to run 45 trials of about 2 minute each. The first 15 under the Minimum Fog condition, the second 15 under the Medium Fog condition, the last 15 under the Maximum Fog condition.

In total, the experiment lasted about 120 minutes (including instructions and breaks between blocks).

As concerning the instructions to the subjects, they were informed about the presence of three different force conditions: one, named Spring Force, in which the stick is felt as a normal joystick (when left, it comes back to the central neutral position). Other two force conditions, named A Force and B Force, in which if the stick is left it moves by itself according to two different laws. After each trial they were asked to recognize the type of force they felt (A, B or Spring Force) trying to define each of them.

4.5.1 Experimental Results

Mean number values of collisions for the three force conditions [NoEF, IHA-OAF, DHA] were entered in a one-way repeated measures analysis of variance (ANOVA). See the results in Figure 4.14.

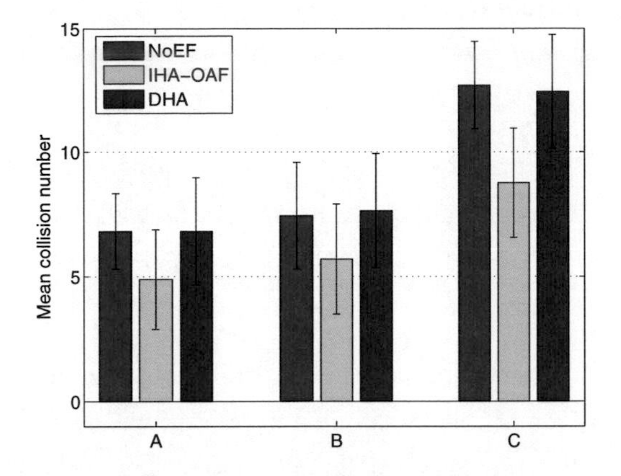

Figure 4.14. Performance (mean and standard deviation) for the 3 Force conditions (DHA, IHA-OAF, NoEF) and for the 3 visibility conditions (A, B, C).

A main effect of the fog condition was found:

$$F(2,9) = 18.366, p < 0.001$$

Post-hoc tests using Bonferroni correction for multiple comparisons, $p < 0.05$ confirmed that the subjects performed significantly worse in the Maximum Fog condition than in the Minimum and in the Medium ones.

A main effect of the force condition was found as well:

$$F(2,9) = 6.427, p < 0.01$$

Post-hoc tests using Bonferroni correction for multiple comparisons, $p < 0.05$ confirmed that the subjects performed significantly better when the IHA-OAF haptic cue was provided in the haptic device than when both DHA and NoEF were provided.

No interaction was found between the two variables.

In other words, the just introduced IHA-Obstacle Avoidance Feel was proved to provide the best results in the obstacles avoidance task irrespective of the fog condition. Thus, the subjects collided less times aided by the IHA-OAF than both the DHA and the NoEF.

This is a pretty surprising result as long as under the Minimum Fog condition the best performance were expected with the NoEF case. According to the present results instead, the employment of IHA-OAF reported the best performance under all the visibility conditions.

Furthermore, better performance of the DHA than the NoEF was expected in presence of both Minimum and Maximum Fog conditions. This seems to be against previous results [42]. A possible explanation is that under both the DHA and the IHA conditions a haptic help (not given in the NoEF case) was given in finding again the main street once lost right after a collision. This help was constituted by the presence of the non null force field inside the obstacle in case of both DHA and IHA. Thus, while in NoEF case was not possible to find again the main street once collided, with both DHA and IHA cases it was easier; to be precise, the best help in finding again the main street was given by the DHA which gave the clearest suggestion about where to go to get out from the collided building because being compliant already helps a lot. A second possible explanation is the different type of baseline condition employed: a difference in the stiffness constant chosen ($120N/m$ of the present work, in Table A.2, against about *200 N/m* of [42]). A third possible explanation is that the DHA force in the present work was weaker than the one employed in previous works and this would make easier to fly close to the obstacles with not too much effort.

After each trial the subjects were asked what kind of force they felt to check if they were able to recognize the type of forces trying to classify them.

Most of them were very able to distinguish between the Spring Force condition (see Section 4.5) and the force feedback conditions (both A Force and B Force). It was, in general, more difficult to classify and distinguish the A and the B Forces.

Some of them correctly noticed and reported the difference between A and B in terms of cue direction with respect to the obstacles (force pushing away from or towards the obstacles). Other participants were only able to identify the difference in strength (actually not present because the amplitude of the force in the two force conditions was exactly the same for the same distance between the aircraft and the obstacles). Someone's classification was really poor (till the end of the 45 trials they still were not able to classify and recognize the force conditions).

Three participants out of 10 were not able to recognize more than the 40% of the forces during the 45 trials.

Only 6 participants out of 10 were able to recognize more than the 60% of the trial forces. Only 3 of them were able to recognize more than about 75% of the same.

After the 45 trials, pilots were interviewed separately. In order to compare the results, each pilot was asked to fill out a questionnaire with 6 questions which is the same as in the CAAF VS DHA Experiment (see Table 3.1 of Section 3.6.4).

The answers to the questionnaire of the 3 only subjects who recognized more than about the 75% of the forces step by step during the 45 trials, are for sure more meaningful than the others (see Figure 4.15).

Figure 4.16 shows instead the answers of the 6 participants able to recognize only the 60% of the trial forces.

It seems that the haptic cues in general (both DHA and IHA-OAF) were retained to be the stronger forces (Questions A) and the forces which produced the most efforts (Questions D) with respect to the NoEF.

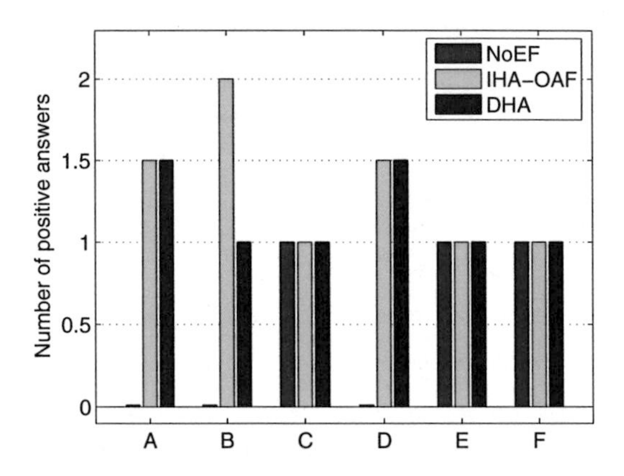

Figure 4.15. Answers to the questionnaire for the 3 participants who recognized more than the 75% of the trial forces.

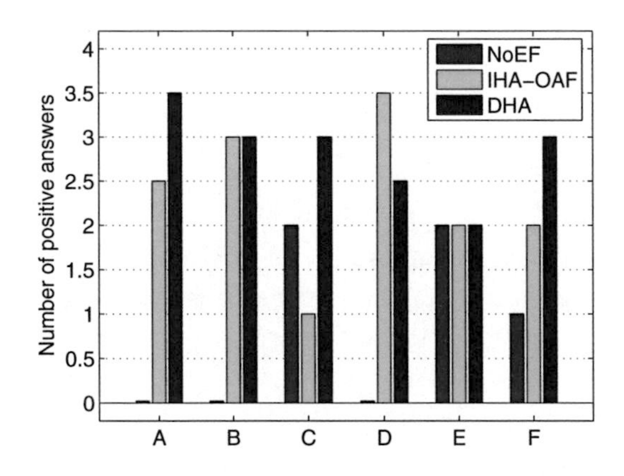

Figure 4.16. Participants answers to questionnaire for the 6 participants who recognized more than the 60% of the trial forces.

But DHA and IHA-OAF were also considered as the most helpful forces (Questions B). Similarly, the NoEF condition was thought to produce no efforts, softer forces but without proving a useful haptic cue (i.e. not helping at all).

About the evaluation of their own performance in the task (Question E), about the condition which gave them the best control on the aircraft (Questions C) and about their own preference between the forces (Questions F) they were more or less divided.

In summary, it was shown that Indirect Haptic Aid could provide better help for subjects than the Direct Haptic Aid and a baseline case (NoEF case, i.e. visual feedback and only the elastic component of the force) in an obstacle avoidance task with a simulated aircraft, confirming the importance to have such haptic feedback in addition to visual feedback to improve the flight security in case of (tele-)operated systems even in pretty good visibility conditions.

From the answers to the questionnaire, it seems that the degree of helpfulness of the haptic cues (both DHA and IHA-OAF) has to be paid through strongest forces feelings and the addition of some effort (the author published the results in [1, 5]). Concerning the IHA force, this seems to be a good compromise to get the best performance!

5 The Mixed CAAF/OAF

This Chapter extends the previously described haptic aid systems by merging them into a system capable of aiding a pilot involved in an obstacle avoidance task in presence of lateral wind gusts.

The simulation environment is the same described in Section 4.1 with the addition of sudden lateral wind gusts.

The remote piloted flight in the presence of environmental constrains is a dangerous task itself as long as a crash of a UAV during teleoperation could not only lead to possible damage to the local environment, but could also lead to the loss of the vehicle followed by the failure of the mission.

Usually UAV missions happen in outdoor environments, thus the UAV is very often subjected to adverse weather conditions. The most dangerous windy condition is represented by the sudden wind gusts that, if not appropriately and suddenly compensated, for example in a constrained mission environment (e.g., a urban canyon) could bring to a fatal collision. As a matter of fact, the buildings of an urban canyon can disturb the airflow creating strong vortices and eddies, tunnel and wake effects (graphically described in Figure 5.1) among other things, which happen in the horizontal plane. For example, in the narrow street "canyons" the wind speed is significantly increased at street corners where lateral streets across the main street and the *tunnel effect* (the only implemented in the following experiments) takes place.

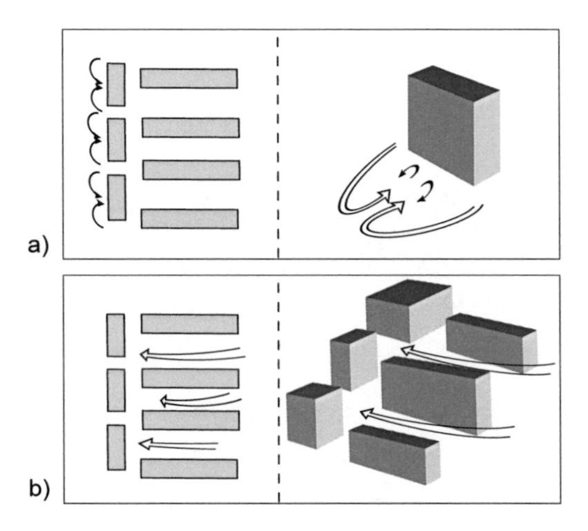

Figure 5.1. The interaction between the wind and the urban canyon: a) the wake effect, b) the tunnel effect.

Two versions of CAAF haptic aid capable of helping to compensate for lateral wind gusts preventing the mission failure will be presented in the first part of this Chapter. It will be shown that designing a new IHA implementation appears straightforward, while designing a DHA system instead can be very complex, especially if the aircraft trajectory is not pre-defined.

After designing both the IHA and DHA to help the pilot in the lateral wind gust rejection, the same force feedback employed in the Chapter 4 will be added to the wind gust haptic aid in order to help the remote pilot in a doubled task: an obstacle avoidance task in a windy environment.

The resulting IHA-based haptic aid is named *Mixed Conventional*

Aircraft Artificial Feel/Obstacle Avoidance Feel and referred as Mixed-CAAF/OAF. It will be shown to definitely improve the pilots' performance with respect to the other approaches (see later) improving the safety of the teleoperation by keeping higher the attention of the pilot in the task (the author published the results in [6]).

5.1 CAAF for lateral dynamics

As concerning, the IHA-based feel for lateral dynamics, thinking about and designing a force expression is very easy. Two examples in the next Subsections are presented: Section 5.1.2 presents the first feedback type that relies on changes of the sideslip angle (it is analogous with what seen in Section 3.3.2) and Section 5.1.3 presents a different approach based on the lateral acceleration produced by the wind gust on the aircraft dynamics.

The Section 5.1.1 explains how the lateral wind gust is simulated.

5.1.1 The Wind Gust Simulation

By hypothesis, only the wind tunnel effect of Figure 5.1 takes place during the simulation. The present Section describes how the tunnel effect is implemented.

As in Chapter 4, the aircraft dynamics is decoupled and only the lateral dynamic is considered (see Section 4.2). The only difference is represented by the addition of the lateral wind gusts that affect the aircraft lateral dynamics.

In both IHA and DHA cases, the lateral wind gust is simulated starting from a triangular velocity profile as wind disturbance: the lateral gust starts at the position $x_{OB} = x_1$ and ends at the position $x_{OB} = x_2 = x_1 + \sim 20m$

(20 meters is the width of the lateral streets) as it happens in the presence of lateral wind tunnels (see Figure 5.1) that cross the main street where the aircraft is flying. The maximum magnitude, in the experiment set to $40kts$, of the wind gust is reached at the position $x_{OB} = (x_1 + x_2)/2$.

For simulating a realistic sense of wind gust, the above described triangular velocity profile is fed to a second order filter which output, v_W, is summed up to the lateral velocity of the aircraft in Earth Reference Frame, $\dot{y}_e$:

$$\dot{y}'_e = \dot{y}_e + v_W \tag{5.1}$$

In Equation (5.1), v_W is the filtered lateral wind gust in Earth Reference Frame, while $\dot{y}_e$ and $\dot{y}'_e$ are the lateral aircraft center of gravity velocity in Earth Reference Frame respectively before and after the lateral wind gust.

Afterwards, the roll angle ϕ, the yaw angle ψ and the aircraft center of gravity velocities in Earth Reference Frame after the wind gust, $\dot{x}_e$ and $\dot{y}'_e$, are employed to calculate the aircraft center of gravity velocities in Body Reference Frame $\dot{x}_B$ and $\dot{y}_B$ through the Equation (5.2):

$$\begin{bmatrix} \dot{x}_B \\ \dot{y}_B \end{bmatrix} = \begin{bmatrix} cos(\psi) & sin(\psi) \\ -sin(\psi)cos(\phi) & cos(\psi)cos(phi) \end{bmatrix} \begin{bmatrix} \dot{x}_e \\ \dot{y}'_e \end{bmatrix} \tag{5.2}$$

Figure 5.2 depicts what just explained.

Equation (5.2) is implemented in the F_{EB} block, and the lateral velocity in Body Reference Frame output ($\dot{y}_B$) is fed to the causal filter $s/(s+100)$ to produce the lateral acceleration $\ddot{y}_B$ as noiseless as possible.

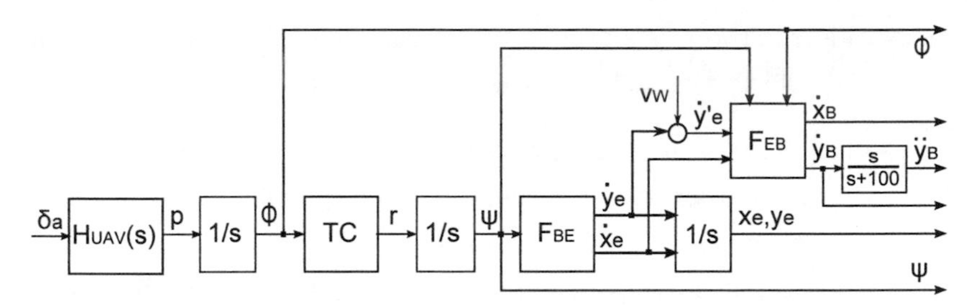

Figure 5.2. The wind gust implementation in the aircraft dynamics.

5.1.2 β-CAAF

As already discussed, UAVs pilots often are manned aircrafts pilots as well, thus they expect in the presence of external disturbances such as wind gusts or turbulences, a cue which is similar to the one they would feel while piloting the aircraft on board. Once more, in order to inform the remote pilot about the external disturbances, an attempt to reproduce, through the haptic feedback, a feeling which mimics the real one is made. The lateral wind gust haptic feedback would produce an immediate effect on the pedals because it affects the rudder which is mechanically commanded through the pedals and this would make the pilot to reject the gust as soon as possible avoiding the consequent changing in the yaw angle which, if not adequately addressed, could bring to dangerous collisions with the environmental constrains.

As in Chapter 4, the rudder deflection δ_r and α (see Section A.1) are supposed to be fixed at the respective trim condition values thus, the only input to the lateral aircraft dynamics (the only one present in this Chapter) is represented by the ailerons commanded through the lateral motion of the haptic device end-effector. This seems to be a reasonable hypothesis as long as an aileron deflection produces first a roll rate and afterwards a yaw rate and, vice versa, the lateral wind gust produces a yaw rate and a roll rate as well. Finally, the hypothesis of only the ailerons as lateral input is justified by the lateral coupling between the rolling and the yawing moments, both created by both an aileron and a rudder deflection.

The lateral wind gust affects above all the sideslip angle β. Due to the previous considerations and given the analogy with the longitudinal dynamics, the force felt on the stick associated to the wind gust in this case is approximately proportional to both the dynamic pressure, q, and to the change of the sideslip angle, β, with respect to its value in trim conditions (see Equation (5.3)) by analogy with the simplified longitudinal force in Equation (3.3).

$$F_{WG,y} \propto q_{trim} \cdot (\beta - \beta_{trim}) \tag{5.3}$$

By following the same considerations made in Section 3.3.2, the lateral proportionality constants are chosen as the longitudinal ones by considering the constrain that one half of the total haptic feedback (heuristically set to $10N$) has to be given by the wind gust aid and the other half by the obstacle avoidance haptic aid. The same maximum velocity as in Section 3.3.2, V_{max} and a maximum value of $40deg$ for the sideslip angle are used.

Equation (5.4) shows how the sideslip angle is computed:

$$\beta = \arctan\left(\frac{\dot{y}_B}{\sqrt{\dot{x}_B^2 + \dot{z}_B^2}}\right) \tag{5.4}$$

Note that, as by hypothesis, the motion and the relative wind are in the horizontal plane only, thus $\dot{z}_B = 0$.

5.1.3 Lateral Acceleration-CAAF

An alternative method for implementing the IHA-based lateral wind gust rejection for more easily comparing IHA and DHA is described now.

The signal chosen to be felt by the pilot through the control device is the lateral acceleration in Body Reference Frame. It is calculated as shown in Figure 5.2.

The obtained lateral acceleration in Body Reference Frame, $\ddot{y}_B$, multiplied by a heuristically chosen constant (in order to obtain a force of about $5N$, that is the 50% of the total haptic aid) creates the feedback for the pilot to inform him/her about the presence of the lateral wind gust.

Thus, in this case:

$$F_{WG,y} \propto \ddot{y}_B \tag{5.5}$$

5.2 Lateral Acceleration-DHA

For comparison purposes, a DHA system is designed using lateral acceleration; the same lateral acceleration of Section 5.1.3, $\ddot{y}_B$, is employed in this case. The lateral acceleration is compared with the zero value and the result is fed to a compensator, in Equation (5.6), which job is rejecting the lateral wind gust through nullifying the lateral acceleration $\ddot{y}_B$. The compensator gain is scaled (of about 80%) in order to require the need of pilot action and to get a maximum haptic feedback value of about $5N$, as in the IHA case.

$$C_{WG}(s) = \frac{F_{WG,y}(s)}{e_{acc}(s)} = \frac{102.0894s + 0.4717}{s + 0.0048} \tag{5.6}$$

e_{acc} is the error between the current lateral acceleration and the zero value. See Appendix C (Section C.2) for details about the compensator of Equation (5.6) design.

5.3 Obstacle Avoidance Force Field

The obstacle avoidance force field for both IHA and DHA simulators is the same as in the Section 4.3.2 with the only difference that here the magnitude is scaled to get an amount of about $5N$.

5.4 Haptic cues for lateral dynamics

Since, by hypothesis, the only dynamics to control is the lateral one, the haptic aid for the obstacle avoidance in windy condition task is applied only to the control device lateral axis (y axis in Figure 3.2), which is thus, the only direction of the force transmitted to the operator.

The total haptic aid F_y needed to run the experiments concerning the obstacle avoidance in windy conditions is shown in Equation (5.7).

$$\begin{cases} F_y = F_{SD,y} + F_{OA,y} + F_{WG,y} \\ F_{SD,y} = F_{S,y} + F_{D,y} \end{cases} \tag{5.7}$$

$F_{S,y}$, $F_{D,y}$ are exactly the same as in Chapter 4. The obstacle avoidance force term, $F_{OA,y}$, depends on the experimental conditions. Three types of external force $F_{OA,y}$ were compared: DHA, IHA and a baseline force condition (see later). The value of $F_{OA,y}$ in both IHA and DHA cases is taken from the Chapter 4 but scaled in magnitude.

The wind gust rejection aid term, $F_{WG,y}$, depends on the experimental conditions as well. The IHA condition value is given in the Equation (5.5) while the DHA condition value is given in the Equation (5.6).

The conditions compared through this experiment are: DHA (both obstacle avoidance and wind gust rejection aids from DHA case), IHA (both obstacle avoidance and wind gust rejection aids from IHA case) and a baseline force condition in which both $F_{OA,y}$ and $F_{WG,y}$ in Equation (5.7) are set to zero.

To create the Indirect and Direct Haptic Aids two simulators are prepared (see Section 5.5): the Mixed-CAAF/OAF (IHA) is compared to the DHA approach through the evaluation experiment of Section 5.6.

5.5 The Windy Obstacle Avoidance Simulators

In order to test the IHA-Mixed CAAF/OAF concept, three simulators are created: the NoEF Simulator, the DHA Simulator and the Mixed-CAAF/OAF (IHA) Simulator.

Subsections 5.5.1, 5.5.2 and 5.5.3 describe the simulators built in order to test the performance in the obstacle avoidance task in the presence of lateral wind gusts object of this Chapter.

5.5.1 NoEF Simulator

Figure 5.3 shows the baseline scheme (i.e. no haptic aids) employed in the obstacle avoidance with wind gusts setup.

Note that the only difference between the present Chapter simulation and the Chapter 4 (Figure 4.2) simulation is the addition of the lateral wind gusts in Figure 5.3 through the v_W signal that represents the gusts in Earth Reference Frame y-axis.

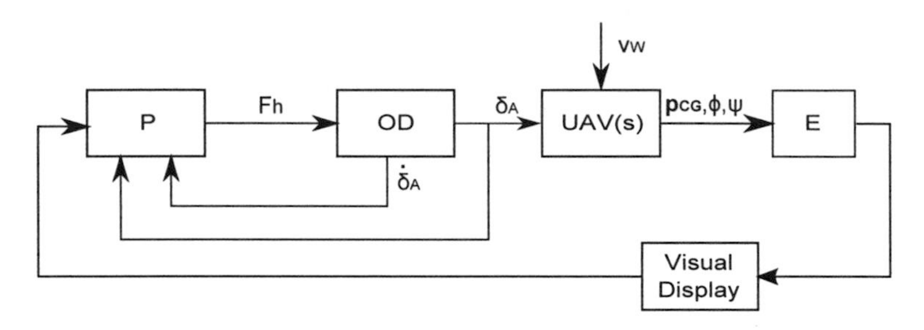

Figure 5.3. The obstacle avoidance with lateral wind gusts simulation baseline scheme.

The aircraft lateral dynamics employed in this Chapter is exactly the same as in Section 4.2 and the same hypothesis (such as coordinated turn) are employed here as well.

This case represents just a visual aid being the haptic feedback only related to the actual stick displacement and to its rate (as in a fly-by-wire like system). In fact, in this case (see Equation (5.7)) $F_{OA,y}$ and $F_{WG,y}$ are set to zero. $F_{SD,y}$ is the same as in the OAF VS DHA Experiment (see Section 4.4).

The pilot has the same virtual scenario employed in Chapter 4 as the only cue of the virtual buildings as seen from the aircraft center of gravity (see Figure 4.1).

Thus, when the wind gusts affects the aircraft, in the case of NoEF feedback, the gust is perceived thanks to the visual feedback only through the sudden variation in the aircraft attitude caused by the lateral wind gust.

The visual feedback is the same in all the conditions of the experiment.

The only haptic feedback felt by the pilot is proportional to δ_A and $\dot{\delta}_A$ produced only from the own input force F_h.

Figure 5.4 depicts an example aircraft trajectory during a simulation trial.

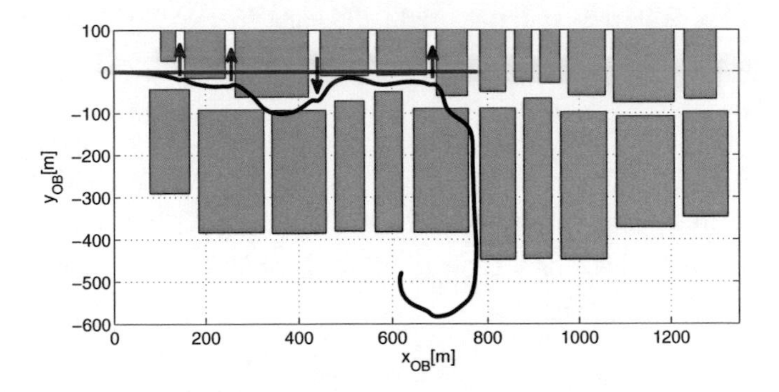

Figure 5.4. NoEF simulation example. The black, the blue and the red lines (superimposed) represent respectively the aircraft trajectory, the obstacle avoidance force (F_{OA}) and the wind gust rejection force (F_{WG}). The black arrows depict the gusts direction.

5.5.2 DHA Simulator

Figure 5.5 shows the block diagram of the simulation system used to test the DHA concept.

As concerning the visual feedback (the same as in NoEF Simulator), the pilot may perceive the distance from the obstacles through the visual display (see Figure 4.1).

While as concerning the haptic feedback, the haptic device is controlled as in Equation (5.7) to behave as a spring-damper system with two additional forces: $F_{OA,y}$ and $F_{WG,y}$. $F_{SD,y}$ and $F_{OA,y}$ are equivalent to the OAF VS DHA Experiment (see Section 4.4).

Thus, an external force composed of two components is given to the pilot through the haptic device: the total force exerted by the obstacles $\mathbf{F_{OBS}}$ (the same as in Section 4.3.2) which is fed into the DHA_{AO} block to output the obstacle avoidance DHA haptic aid $F_{AO,y}$ and the wind gust rejection aiding force $F_{WG,y}$ produced by the compensator represented by the DHA_{WG} block (see Figure 5.5) as explained in Section 5.4.

This compensator is added to help the pilot in rejecting the lateral wind gusts being designed in order to cancel the lateral acceleration $\ddot{y}_B$ produced by them.

Both the haptic cues, $F_{OA,y}$ and $F_{WG,y}$ (F_{OA} and F_{WG} in Figure), together with the pilot force input F_h, are fed to the haptic device (OD block in Figure 5.5) to produce the stick deflection δ_A. The δ_A and $\dot{\delta}_A$ feedback indicate the proprioceptive feedback.

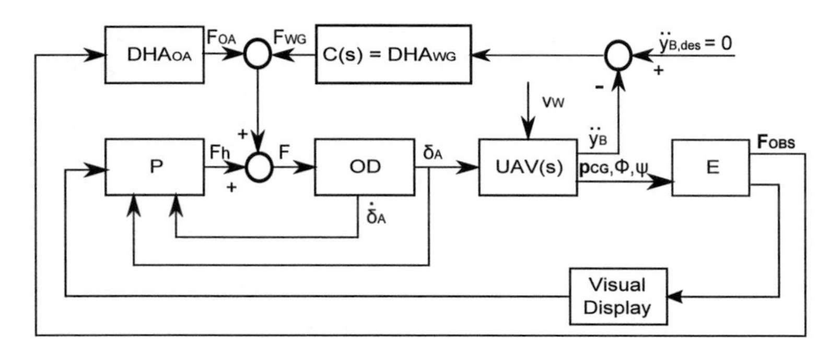

Figure 5.5. DHA-based obstacle avoidance in the presence of lateral wind gusts simulator scheme. The haptic forces F_{OA} and F_{WG} deflect the stick inducing a helpful change of the aircraft trajectory.

The obstacle avoidance feel, part of the current haptic feedback, works exactly as the one of Chapter 4: again, $F_{OA} = F_{B,y}$ in the Equation (4.9).

Suppose the aircraft is affected by a lateral wind gust, a lateral acceleration in Body Reference Frame rises, the compensator detects it and produces a force which would, at least partially, make it null. In fact, the pilot should follow and amplify it (remember that the gain of the compensator is scaled by the 80%) in order to make the lateral acceleration null, that is to fully reject the lateral wind gust.

Figure 5.6 depicts an example of the aircraft trajectory during a simulation trial.

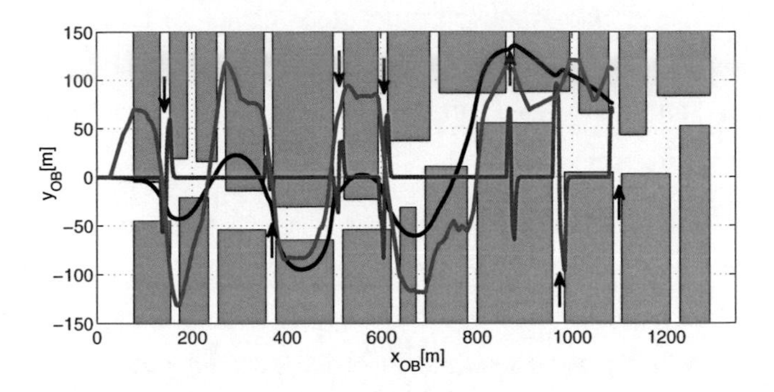

Figure 5.6. DHA simulation example. The black, the blue and the red lines represent respectively the aircraft trajectory, the obstacle avoidance force ($F_{OA} * 10[N]$) and the wind gust rejection force ($F_{WG} * 10[N]$). The black arrows depict the gusts direction.

5.5.3 IHA-Mixed CAAF/OAF Simulator

By following the same principle as in all the previously described IHA-based haptic feedbacks, the Mixed-CAAF/OAF should produce a "natural" force sensation when both an obstacle is approaching (see Chapter 4) and a lateral wind gust is affecting the aircraft.

For example, if the wind gust comes from the right side of the aircraft, the lateral acceleration of the aircraft will increase towards the left and also the stick would move towards the left. Again, the pilot would naturally oppose this movement by rolling on the right (stick on the right), that is the direction needed to reduce the lateral acceleration generated by the gust.

Once more, the stick moves through a shifting of the stick neutral point.

The vanishing of the haptic cue informs the pilot that the gust vanished as well.

As concerning the visual feedback (the same as in both NoEF and DHA Simulators), the pilot may perceive the distance from the obstacles through the visual display (see Figure 4.1).

Regarding the haptic feedback, the haptic device is controlled as in Equation (5.7) to behave as a spring-damper system with two additional forces: $F_{OA,y}$ and $F_{WG,y}$ (F_{OA} and F_{WG} in Figure 5.7) as explained in Section 5.4.

Thus, a force with two components is given to the pilot through the haptic device: the total force exerted by the obstacles $\mathbf{F_{OBS}}$ (the same as in Section 4.3.2) which is fed to the IHA_{AO} block to output the obstacle avoidance IHA haptic aid $F_{AO,y}$ and the wind gust rejection aiding force $F_{WG,y}$ produced by the IHA_{WG} block (see Figure 5.7). This force just transmits to the pilot the lateral acceleration produced by the lateral wind gust.

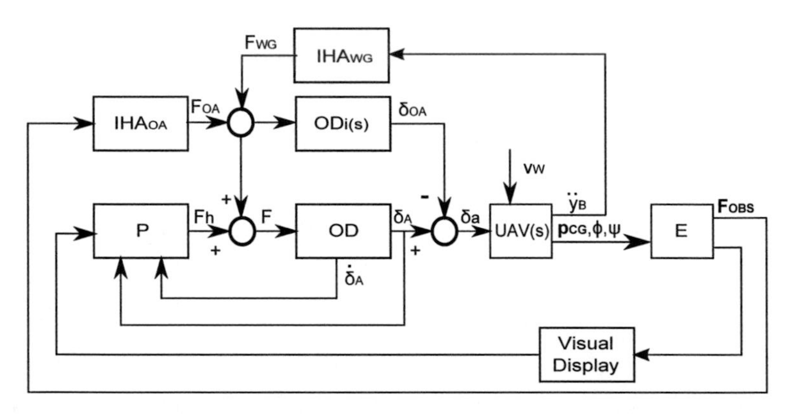

Figure 5.7. IHA-Mixed CAAF/OAF simulator scheme. The haptic forces F_{OA} and F_{WG} deflect the stick without producing any change to the aircraft trajectory thanks to the effect of the compensating signal δ_{OA}.

As concerning the wind gust rejection feel in Mixed-CAAF/OAF case, an example is illustrated: suppose the aircraft is affected by a lateral wind gust, a lateral acceleration in Body Reference Frame arises; the pilot would naturally oppose it by rejecting the wind gust and, as a consequence, will hopefully avoid a potential collision that might occur in case the wind gust is not readily and suddenly rejected.

Thus, the pilot will feel a haptic information about both the presence of the obstacle and the presence of a lateral wind gust and he/she will see it through the visual display (when the visibility condition is good enough, i.e. no foggy weather).

Figure 5.8 depicts an example of the aircraft trajectory during a simulation trial.

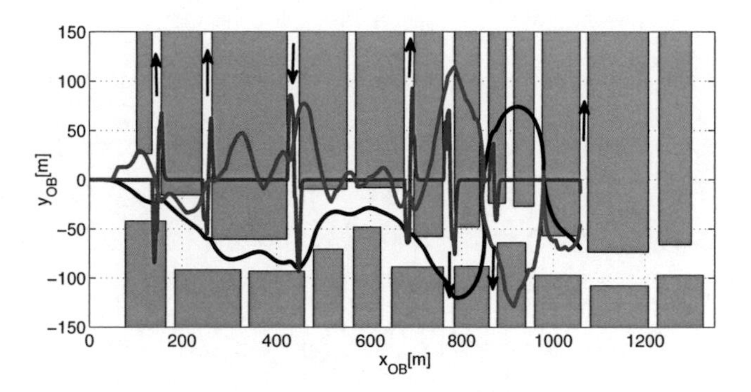

Figure 5.8. IHA-Mixed CAAF/OAF simulation example. The black, the blue and the red lines represents respectively the aircraft trajectory, the obstacle avoidance force ($F_{OA} * 10[N]$) and the wind gust rejection force ($F_{WG} * 10[N]$).The black arrows depict the gusts direction.

In other words, the IHA-Mixed CAAF/OAF follows the general IHA concept described before: it supplies to the pilot the information about the presence of the obstacle on a side of the aircraft and about a lateral wind gust but it does not effect in any way the commands actually sent to the aircraft; this helps the pilot indirectly by improving his/her SA, that is to let him/her know that in the remote environment a collision is going to happen and/or a lateral wind gust is affecting the aircraft and leaving him/her the full authority to take control decisions by changing the vehicle direction of the motion.

The mathematical proof of the concepts described above is similar to the one presented in the Subsection 4.4.3 with the final result that F_{OA} and F_{WG} change just the Omega Device neutral point of δ_{OA} and the only input to the aircraft dynamics is given by the pilot command F_h (Equation (4.17)).

5.6 Mixed CAAF/OAF Evaluation

In order to test the three haptic aiding systems, several experiments of obstacle avoidance in the presence of sudden lateral wind gusts were run.

The present task is even more difficult than the one in Chapter 4 because the street is a bit tighter and also 8 lateral wind gusts (4 toward left, 4 toward right), which exact position was strategically set in each of the five employed scenarios (the same as in previous Chapter), were added.

An attempt to make the experiment as realistic as possible was made, in the sense that the gusts were added where some of the lateral smaller streets cross the main street. The lateral street which is the ideal candidate to host the lateral wind gust is a street in which the physical characteristic might bring to tight turns very close to the buildings to avoid, making the potential collision very likely to happen.

The experiments were run under two different windy conditions, No Wind (*NW*) and Wind (*W*); two different visibility conditions, *Minimum Fog* (same as the first fog condition in the previous Chapter) and *Maximum Fog* (same as the worse fog condition in the previous Chapter), and under three different force condition: DHA, IHA and NoEF. In total the conditions were twelve.

Note that the worse visibility condition, shown in Figure 4.13c, is even more dangerous in windy conditions than it was in the previous Chapter.

Thus, a even stronger effect about the performance is expected (again the number of collision is chosen as metric).

The experimental task is the same as in the previous Chapter: to get the end of the street by avoiding the collisions with buildings although the presence of 8 lateral wind gusts. Again, to test the natural response to the different types of force no instructions were given to the participants about the force they were going to feel on the stick.

The goal of these tests is to prove whether adding the Mixed-CAAF/OAF kinesthetic (force) cue to the visual cue improves the control with respect to the other two conditions. In particular the goal is to assess as analytically as possible the differences in pilot performance in the three cases (NoEF, IHA and DHA). Thus, the performance of the subjects (dependent variable) was measured through the number of collisions in the flight across a constrained environment and in the presence of lateral wind gusts.

Seven naïve subjects participated to the experiment. All had normal or corrected-to-normal vision. They were paid, naïve as to the purpose of the study, and gave their informed consent. The experiments were approved by the Ethics Committee of the University Clinic of Tübingen, and conformed with the 1964 Declaration of Helsinki.

All the trials (see Appendix D for details) were mixed and counter-balanced and no instructions were given about the three different force conditions to test natural reaction of the subjects to the different twelve conditions.

Each fog condition was run as a separate block and counterbalanced as well.

The participants had to run 60 trials of about 2 minute each.

In total, the experiment lasted about 150 minutes (including instructions and breaks between blocks).

As concerning the instructions to the subjects, they were informed about the presence of three different force conditions: one, named Spring Force, in which the stick is felt as a normal joystick (when left, it comes back to the central neutral position). Other two force conditions, named A Force and B Force, in which if the stick is left it moves by itself according to two different laws. After each trial they were asked to recognize the type of force they felt (A, B or Spring Force) trying to define each of them.

After each of the 4 blocks (Wind plus Maximum Fog, Wind plus Minimum Fog, No Wind with Maximum Fog, No Wind with Minimum Fog) and after the whole experiment they were interviewed separately. In order to compare the results, each pilot was asked to fill out a questionnaire with 6 questions (the same questionnaire as in the previous experiments, see the Table 3.1).

5.6.1 Experimental Results

Mean number values of collisions for each of the twelve conditions were entered in a one-way repeated measures analysis of variance (ANOVA). See the results in Figure 5.9.

A main effect of the wind condition was found:

$$F(1,6) = 6.6365, p < 0.05$$

Post-hoc tests using Bonferroni correction for multiple comparisons, $p < 0.01$ confirmed that the subjects performed significantly worse in the Wind condition than in the No Wind.

A main effect of the fog condition was found:

$$F(1,6) = 19.252, p < 0.01$$

Post-hoc tests using Bonferroni correction for multiple comparisons, $p <$

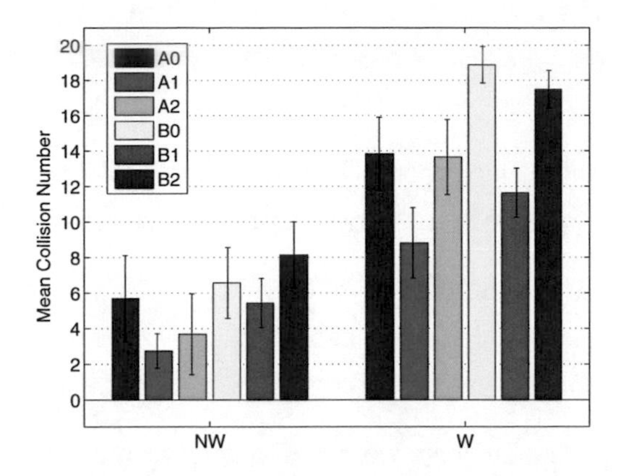

Figure 5.9. Performance (mean and standard error) for the two Wind conditions (No Wind and Wind), for the three Force conditions (DHA=2, IHA-Mixed CAAF/OAF=1, NoEF=0) and for the two visibility conditions (A, B).

0.001 confirmed that the subjects performed significantly worse in the Maximum Fog condition than in the Minimum one.

A main effect of the force condition was found as well:

$$F(2,12) = 16.928, p < 0.001$$

Post-hoc tests using Bonferroni correction for multiple comparisons, $p < 0.05$ confirmed that the subjects performed significantly better when the IHA-Mixed CAAF/OAF haptic cue was provided in the haptic device than when both DHA and NoEF were provided.

No interaction was found between the two variables.

In other words, the just introduced IHA-based Mixed-CAAF/OAF was proved to provide the best results in the obstacles avoidance in windy conditions task irrespective of the fog condition and of the wind conditions. Thus, the subjects collided less times aided by the IHA-based Mixed-CAAF/OAF than both the DHA and the NoEF cases. It is possible to conclude that the employment of IHA-based Mixed-CAAF/OAF improves the performance in obstacle avoidance in all the visibility conditions with and without wind.

Once again, the same observations as in Section 4.5.1 can be made here about the surprising results and the possible explanations.

In particular, Figure 5.4 clearly shows what does getting lost after a collision mean and how in NoEF case is not easy to find again the main street once collided as opposed to both DHA and IHA cases.

After each trial the subjects were asked what kind of force they felt to check if they could recognize the type of forces trying to classify them.

Most of them were very able to distinguish between the Spring Force condition and the force feedback conditions (both A Force and B Force). It was, in general, more difficult to classify and distinguish the A and the B Forces.

Some of them correctly noticed and reported the difference between A and B in terms of cue direction with respect to the obstacles (force pushing away from or towards the obstacles).

Other subjects were only able to identify the difference in strength (actually not present because the amplitude of the force in the two force conditions was exactly the same for the same distance between the aircraft and the obstacles). Someone's classification was really poor (till the end of the 60 trials they still were not able to classify and recognize the force conditions).

Only 4 subjects over 7 were able to recognize more than the 60% of the trial forces. Only 2 of them were able to recognize more than about 70% of the same.

After the 60 trials, pilots were interviewed separately. In order to compare the results, each pilot was asked to fill out a questionnaire (the same as in the previous experiments in Table 3.1).

The answers to the questionnaire of the 2 only subjects who recognized more than about the 70% of the forces step by step during the 60 trials, are for sure more meaningful than the others (see Figure 5.10).

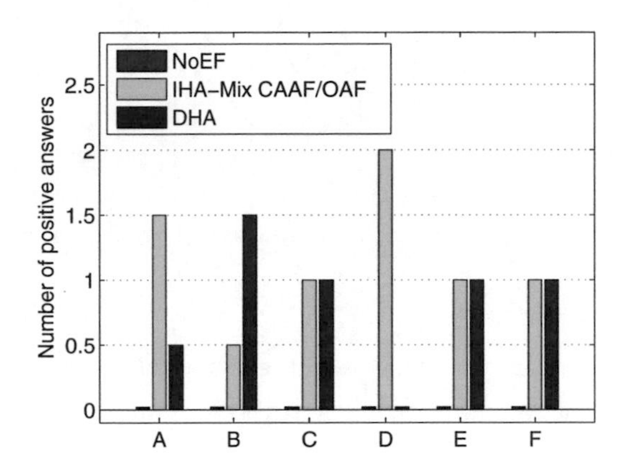

Figure 5.10. Answers to questionnaire for the 2 participants who recognized more than the 70% of the trial forces.

Figure 5.11 shows instead the answers of the 4 subjects able to recognize only the 60% of the trial forces.

It seems that the IHA-Mixed CAAF/OAF in general was retained to be the strongest force (Questions A) and the forces which produced the most efforts (Questions D) with respect to both the DHA and the NoEF conditions. The DHA was considered as the most helpful force (Questions B). As concerning the NoEF condition, it was thought to produce no efforts, weaker forces but without proving a useful haptic cue (i.e. not helping at all).

About the evaluation of their own performance in the task (Question E), about the condition which gave them the best control on the aircraft

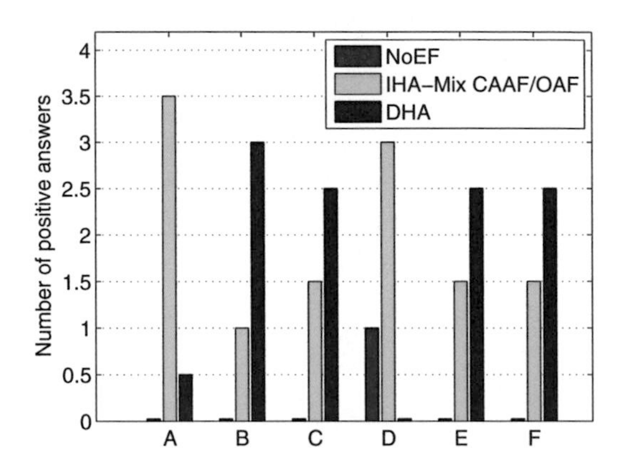

Figure 5.11. Answers to questionnaire for the 4 participants who recognized more than the 60% of the trial forces.

(Questions C) and about their own preference between the forces (Questions F) they were more or less divided between IHA and DHA forces.

What just mentioned are the general results, e.i. the results coming from the final questionnaire regarding all the wind and fog conditions. They are more or less representative of the results which come from the questionnaire after each of the four blocks (No Wind-Minimum Fog, No Wind-Maximum Fog, Wind-Minimum Fog, Wind-Maximum Fog), but an exception has to be reported: when the experimental condition got worse (No Wind-Maximum Fog, Wind-Minimum Fog, Wind-Maximum Fog), the NoEF condition was the conditions the most subjects preferred; in fact, they classified it as most helpful, they felt to have the best control on the aircraft and they thought they obtained with it the best results. This is due maybe to the fact that the worse are the visibility and the windy conditions, the less the participants trusted in the haptic cues (both DHA and IHA) maybe because not enough trained on it.

In summary, the aim of the obstacle avoidance in windy conditions haptic cues evaluation experiment was to test whether the employment of a newly developed IHA-Mixed CAAF/OAF (Obstacle Avoidance Feel) could produce some improvement with respect to other approaches present in literature. It was shown that Indirect Haptic Aid could provide better help for subjects than the Direct Haptic Aid and a baseline case (NoEF case, i.e. visual feedback and only the elastic and damping components of the force) in an obstacle avoidance task in windy conditions with a simulated aircraft, confirming the importance to have such haptic feedback in addition to visual feedback to improve the flight safety in case of (tele-)operated systems even in pretty good visibility conditions.

It seems, finally, that the degree of helpfulness of the haptic cue IHA-Mixed CAAF/OAF has to be paid through strongest forces feelings and the addition of some effort (the author published the results in [6]). This seems to be a good compromise to get the best performance!

6 Delayed Bilateral Teleoperation

A teleoperation system in presence of force feedback is often referred to as a bilateral system in which the human operator controls a remotely located robot.

In the particular case of UAV bilateral teleoperation, the remote pilot is responsible for the UAV at all times and it is crucial that he/she at all times can understand the state of the airborne UAV.

The introduction of a haptic feedback in the UAV's GCS, seems to improve the SA of the remote pilot and hopefully the performance as well.

The haptic feedback introduces a supplementary control loop which acts directly on the pilot control device. The further addition of communication time delays could easily bring the system to instability besides the degradation of the remote pilot SA.

All these troubles could have an affect on system performance and overall safety.

The scientific literature suggests a lot of remedies to solve the instability problems in a delayed bilateral teleoperation system (see Section 2.4). A widely employed method is, for example, the scattering theory through the wave variables approach [8, 52, 63]. That implementation does not seem to be suitable for the present work (see Section 6.3).

A less employed method is represented, instead, by the admittance control [56, 57]: it was shown to improve the stability characteristics of the delayed bilateral teleoperation system at the cost of transparency [56]. Usually, in remote/virtual environment applications, a good transparency/virtual presence is needed when the real remote/virtual environment is asked to be scanned in details (e.g. exploration, manipulation of objects, etc) and for this reason, master/slave/human operator dynamics need to be canceled. In the experimental activities of the present work, a good transparency property it is not really requested; on the contrary the stick dynamics has to be felt by the human operators to obtain good performance as the author showed (see Chapter 3) and published [2] because the pilots use to feel the stick dynamics properties. In the present research the transparency property is simply unwanted.

Usually the admittance control architecture is employed with admittance-type devices. The haptic device employed in this work is an impedance-type device characterized by very light-weight construction with low inertia and friction. However, high dynamic properties (i.e. high stiffness), usually characterizing the admittance-type devices, were chosen in order to simulate a FBW stick-type.

All these reasons have brought to the implementation of the admittance controller in the delayed teleoperation system object of this Chapter.

The admittance controller needs a force sensor in the implementation. This problem is overcome through the design of an observer for the human force. It is proved to work pretty well in simulation and the implementation of the admittance controller plus the observer of the human force will be shown to improve the stability properties of the system under consideration (see later).

The only type of force feedback employed in this Chapter is the DHA-based one designed for the obstacle avoidance task (see Chapter 4).

The Chapter is organized as follows: Section 6.1 describes the bilateral system setup; Section 6.2 depicts the baseline scheme engaged and shows the instability dynamics due to the addition of time delay in the communication link; Section 6.3 reports the implementation of wave variables usually employed to tackle such instability problems and shows that and why they are not suitable for this case; Section 6.4 details the implementation of an admittance controller and proposes the introduction of an observer of the force exerted by the human on the stick showing that the resulting system is capable of standing substantial communication delays (the author published the results in [7]).

6.1 System Setup

This Chapter presents a bilateral teleoperation system in which a human operator, a pilot, controls a remotely located UAV, the slave, via a man-machine interface, the master device, while receiving haptic feedback of the interaction between the UAV and the remote environment, see Figure 6.1.

In details:

- **Master**: the master device task is to simulate a control stick through the use of a high precision force feedback device (omega.3, Force Dimension, Switzerland) (see Section A.2 for details).
- **Slave**: the slave system is constituted by the dynamics of the aircraft

Figure 6.1. The teleoperation system (by http://www.flickr.com). The red arrow over the control device represents the force feedback.

under control; in order to maximize the pilot attention on its task, only the lateral aircraft dynamics is considered: the slave input is the aileron deflection and its output is the aircraft position, heading and heading rate.

- **Environment**: a virtual environment is displayed during the experiments to produce the visual cues; a subjective view from the aircraft cockpit is simulated using a realistic virtual environment created using the DynaWORLDS software package [58]. The environment is the same as in Chapters 4 and 5 (see Section A.3 for details). Figure 4.1 depicts the employed setup.

The just mentioned components of the setup are described in details in Section 6.2.

The control task is the same as in Chapter 4 (narrow street scenario) with the addition of time delays in the communication link. In the design phase, for the purpose of a more straightforward understanding of the system behavior in presence of time delay, a simple scenario with two long obstacles is employed; it represents a straight narrow street, namely a corridor. The aircraft has to be flown in this corridor getting to the end of the street by avoiding the collisions with the virtual buildings. A repulsive force field is associated to the obstacles and it is sent back to the operator through the communication link.

As anticipated, only the DHA approach is tested in the teleoperation environment. Since the DHA approach must produce stick motions that induce beneficial trajectory variations of the aircraft, the DHA system is designed as it would be done with a compensator: a control system that regulates the distance from the obstacles, or equivalently, brings the aircraft to the minimum of the repulsive force field. The total compensator effect is assumed to substitute both human pilot and haptic augmentation system (see later). In order to design it and evaluate its performance, simulations with the pilot out of the loop are performed first (only the DHA compensator acts on the stick) with the simplified simple corridor scenario; afterwards, the system is tested with pilots in the loop and with the non Manhattan-like narrow street scenario of Chapter 4 (see Figure D.3).

As concerning the baseline bilateral scheme implemented in this Chapter, the scheme of Figure 4.8 is thus modified by employing a compensator in place of the human pilot and by implementing a local controller in the slave side. Section 6.2 explains in details the just mentioned scheme.

6.2 F-P scheme

For the purposes of this work a two-channel architecture [44] is employed and two physical signals are exchanged between master and slave: a position command (stick position that encodes the yaw rate command) is sent from the master side to the slave side and a force signal is sent from the slave side to the master side. This scheme is known as Force-Position architecture [44].

The teleoperation scheme considered in this Chapter is schematically illustrated in Figure 6.2.

The classical teleoperation schemes employ a local controller in both the master and the slave side. In analogy to this, a yaw rate command is used as input for the slave side and a local controller is employed to regulate the actual aircraft yaw rate signal to the desired one (see below).

As concerning the master side, since the haptic device used for the experiments (the Omega Device) does not possess a force sensor, an open-loop force control is adopted in the master side and a local controller $C_s(s)$ is instead employed in the slave side.

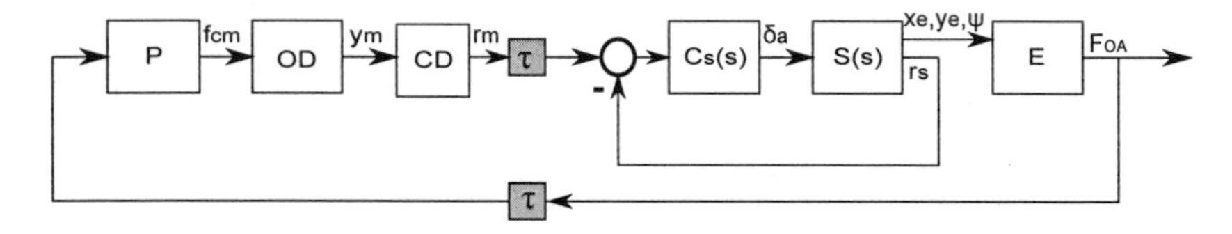

Figure 6.2. The baseline Force-Position scheme.

In Figure 6.2, OD is the Omega Device (the master haptic device); $S(s)$ and $C_s(s)$ are the aircraft dynamics and its local controller. The P block represents the real pilot who produces the force f_{cm} to directly act on the Omega Device providing a displacement, y_m, of the end-effector. This displacement is then converted, through the car-driving metaphor (CD block) [71] (see Section 6.2.1), to a heading rate command used as a reference command, r_m, for the aircraft heading rate, r_s. In order to simulate the system with the real pilot out of the loop a compensator $C(s)$ is designed. In Figure 6.2, the compensator $C(s)$ would take the place of the pilot P. This architecture is chosen with the possibility in mind of future splitting the compensator action in two components as will be detailed and better described later (Section 6.2.7). The local controller $C_s(s)$ is designed (see Section 6.2.3) for the slave side in order to regulate r_s to r_m. The aircraft position (x_e,y_e) and its heading (ψ) are used by the environment block, E, to calculate the force F_{OA}, based on the relative distance between the aircraft and the obstacles, to generate the haptic force for the master side (see Section 6.2.4). τ is the time delay when present. The compensator $C(s)$ is designed (see Section 6.2.6) by making use of the identified model of the Omega Device (Equation (B.2) in Appendix B) to take the feedback force F_{OA} as input and to produce the force input for the Haptic Interface, f_{cm}.

This will be considered as the baseline scheme of this Chapter.

The master and slave dynamics together with their input and output will be explained respectively in the Sections 6.2.5 and 6.2.2.

6.2.1 The Car-Driving Metaphor

A car-driving metaphor [71, 20, 33] for direct control of the UAV is employed. According to it the operator uses the end-effector of the haptic device to designate the desired speed and rate of turn. A logical point (x,y) (obtained by projecting the 3D haptic end-effector location to a xy-plane) is mapped to motion parameters such as speed and turning rate as in Figure 6.3.

A constant longitudinal velocity is chosen for the UAV (see Section 4.2), then only the lateral motion of the end-effector is considered and converted into a heading rate command (in rad) used as a reference command, r_m, for the aircraft heading rate, r_s. Equation (6.1) shows the function implemented inside the CD block of Figure 6.2.

$$r_m = CD(y_m) = 4.380 \cdot y_m \tag{6.1}$$

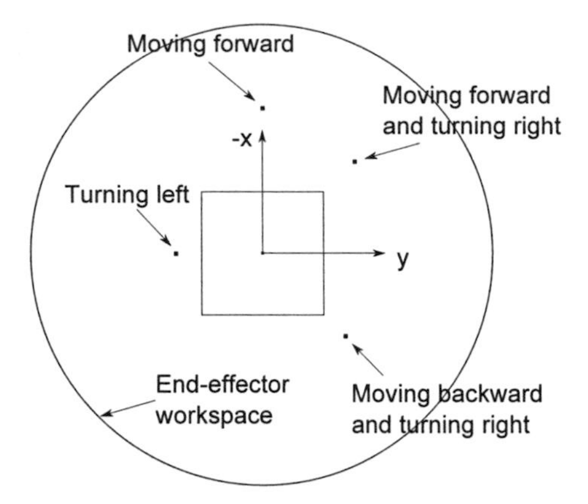

Figure 6.3. Car-driving metaphor: mapping a logical point (x, y) to motion parameters (speed rate, turning rate).

6.2.2 The slave dynamics

The input of the aircraft system $S(s)$ is the aileron deflection, δ_a, that is the output of the $C_s(s)$ controller (see Figure 6.2) and the outputs are the aircraft yaw rate (r_s), position (x_e and y_e) and heading (ψ). The block $S(s)$ of Figure 6.2 is shown in details in Figure 6.4.

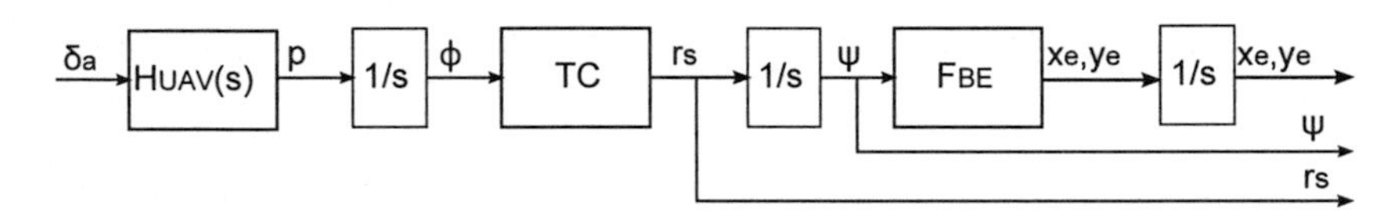

Figure 6.4. The aircraft lateral dynamics.

In Figure 6.4 the transfer function $H_{UAV}(s)$ from aileron, δ_a, to roll rate, p or $\dot{\phi}$, of Equation (6.2) is employed. It is obtained through linearization and dominant poles approximation of the non linear Beaver DHC-2 of the *Flight Dynamics and Control Toolbox* [68]. The roll angle, ϕ, is obtained via integration.

$$H_{UAV}(s) = \frac{-3.7972}{s^2 + 6.9828s + 0.4297} \tag{6.2}$$

As in Chapters 4 and 5, the assumption of the aircraft performing coordinated turns (zero velocity in the lateral body axis, [69]) is made (see Equation (6.3)). The coordinated turn is depicted through the TC block in Figure 6.4) and performed at constant speed (V); the heading rate r_s or $\dot{\psi}$ is calculated through the Equation (6.3):

$$r_s = tan(\phi)\frac{g}{V} \tag{6.3}$$

The rest (i.e. calculation of the heading angle, ψ, and of the aircraft center of gravity coordinates in Earth Reference Frame) is the same as in Section 4.2.

As seen in Figure 6.2, the roll rate r_s is used to calculate the error for the slave controller, while position and its heading are used by the environment block, E, to calculate the force F_{OA}, based on the relative distance between the aircraft and the obstacles, to send back to the master side (see Section 6.2.4).

6.2.3 The slave controller

The slave controller, Equation (6.4), is designed in the linear domain using the Evans' Root Locus tool in order to have a reasonable response time ($1.2s$), a well damped behavior (damping factor of about 0.9) and a limited motion for the aileron surfaces (less than 50% of maximum aileron dfelection).

$$C_s(s) = \frac{-7,2672s - 3,6336}{0,17s + 1} \tag{6.4}$$

Figure 6.5 shows the Root Locus plot used for the design. See depicted the open loop poles as blue diamonds, the compensator roots in red and the closed loop poles as green squares.

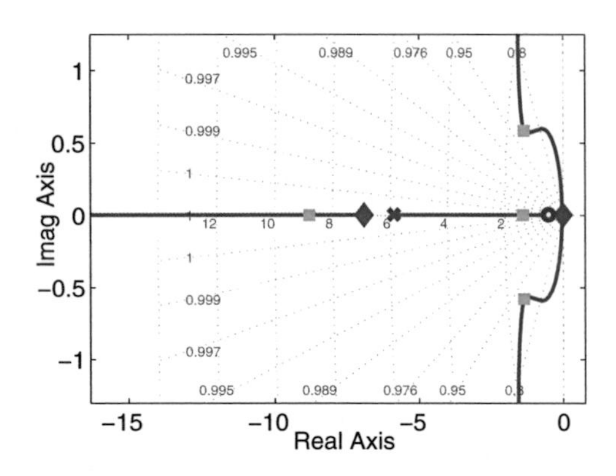

Figure 6.5. The slave root locus used to design the compensator $C_s(s)$.

6.2.4 The haptic feedback

As in Chapters 4 and 5, the only aircraft dynamics to be controlled is the lateral one and the haptic aid for the obstacle avoidance task will be only in the lateral axis of the stick (actually the Omega Device), that is the y axis in Figure 3.2.

A system where the haptic interface appears as a stick with constant damping and stiffness with the addition of an external force coming out when needed (namely when near obstacles) is designed. Then, the force $F_{S,y}$ felt by the operator during the obstacle avoidance task is the same as in Equation (4.6).

The force field around the obstacles (again in the fixed Earth Reference Frame) is the same as in Equation (4.7) of Section 4.3.2.

As concerning the force field generated by a single obstacle a small difference is now introduced: in order to simplify the force field in which the aircraft flies, a different versor than the one used in Equation (4.8) is chosen.

In fact, the unity vector $\frac{\mathbf{p_{OB}}-\mathbf{p_{CG}}}{\|\mathbf{p_{OB}}-\mathbf{p_{CG}}\|}$ is now employed (Equation (6.5)). The meaning of the symbols is the same as in Section 4.3.2. The force field is aligned with the vector distance between the aircraft center of gravity and the obstacle; thus, the force field is always perpendicular to the obstacles' walls (in the obstacles' vertices it is radial but here it is not relevant as long as the simulations take place along the obstacles' sides as in the corridor scenario; see later).

Figure 6.6 shows an example of the force field produced by the obstacles. Module and direction of the force field at the current position of the aircraft are used in the simulator to generate the haptic sensation.

As in Section 4.3.2, the total force exerted by the obstacles is expressed in the fixed Earth Reference Frame and a change in the aircraft Body Reference Frame is necessary (see Equation (4.9)).

$$\mathbf{F_{OB}} = \begin{cases} -k_{OA} \cdot (d(\mathbf{p_{OB}}, \mathbf{p_{CG}}) - r_e) \cdot \frac{\mathbf{p_{OB}}-\mathbf{p_{CG}}}{\|\mathbf{p_{OB}}-\mathbf{p_{CG}}\|}, & \\ & \textit{for}\; d(\mathbf{p_{OB}}, \mathbf{p_{CG}}) < r_e \\ 0, & \textit{otherwise} \end{cases} \tag{6.5}$$

The distance between the obstacles (namely the width of the corridor) is set to $2r_e$ (see Section 4.3.2), then the force field has a *V shape* (see

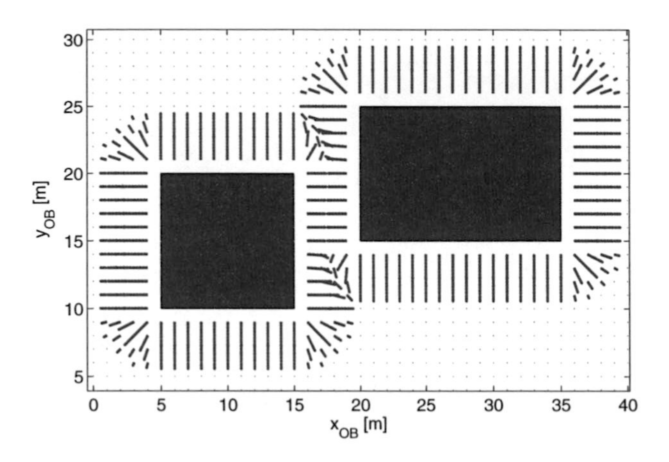

Figure 6.6. Example of the obstacle repulsive force field.

Figure 6.7) with null force in the middle of the corridor and the maximum force (about $8N$) at the obstacles sides. Note that the haptic force F_{OA} is proportional to the distance of the aircraft from the middle of the corridor.

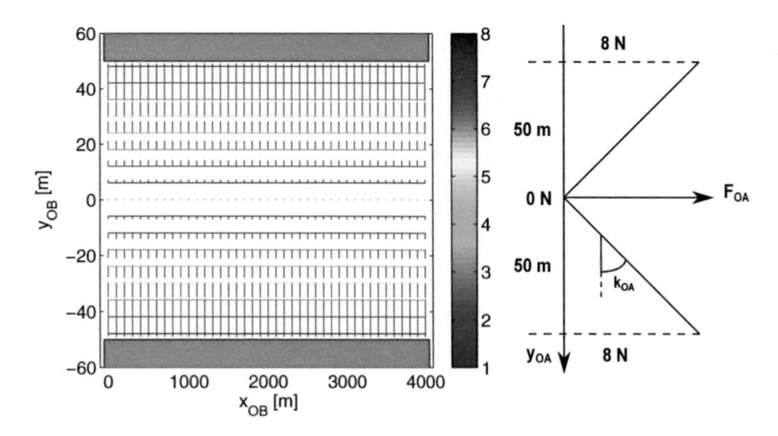

Figure 6.7. Corridor repulsive force field with contour lines. Note the null force field in the middle of the corridor.

Assuming that the reference frame where the position of the aircraft is defined has its $\mathbf{x}$ axis aligned with the corridor and its origin in the middle of the corridor, the force field can be hypothesized:

$$F_{OA} \cong k_{OA} \cdot y_{OA} \tag{6.6}$$

and y_{OA} assumes zero value, thus producing zero force, in the middle of the corridor. The corridor generated force field inclusive of the contour lines is depicted in Figure 6.7.

6.2.5 Omega Device dynamic model

Haptic devices are usually modeled as a simple mass (M), thus their transfer function is usually:

$$\frac{1}{Ms^2}$$

As anticipated above, a system where the haptic interface appears as a stick with constant damping and stiffness with the addition of an external force is designed. Thus, the stick transfer function would be:

$$\frac{1}{Ms^2 + Bs + K}$$

Due to its non-idealities (friction, actuator dynamics etc.) the Omega Device actual behavior, with the added stiffness and damping, has to be identified (see Section B for details).

The transfer function $OD_y(s)$ obtained is shown in Equation (B.2).

6.2.6 Master Compensator Design

The compensator of Equation (6.7) is designed in the linear domain using the Evans' Root Locus tool, in order to have a reasonable response time (about $4s$), a well damped behavior (damping factor of about 0.5) and a limited force for the Omega Device (about $4N$).

$$C(s) = \frac{2.799s + 0.8748}{s + 10} \tag{6.7}$$

The compensator $C(s)$ produces a force on the stick that acts as the sum of the human operator force and the haptic aid itself.

Figure 6.8 shows the root locus used for the design. See depicted the open loop poles as blue diamonds, the open loop zeros as blue circles, the compensator roots (poles as crosses, zeros as circles) in red and the closed loop poles as green squares.

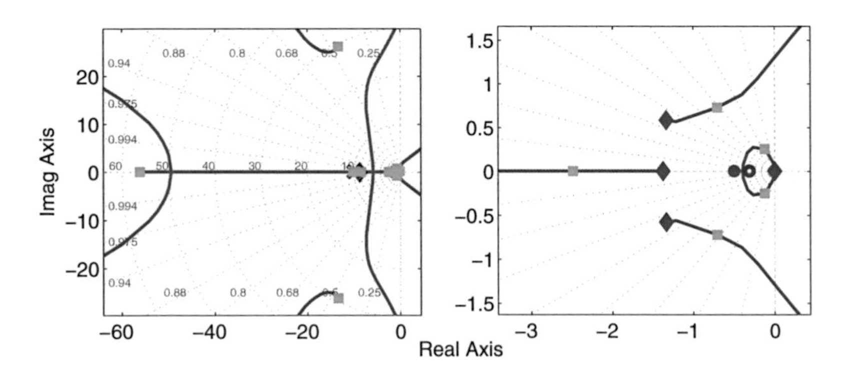

Figure 6.8. The system root locus to design the compensator $C(s)$. On the right side is shown a zoom in on origin.

Figure 6.9 depicts the Bode plot of the compensator in Equation (6.7).

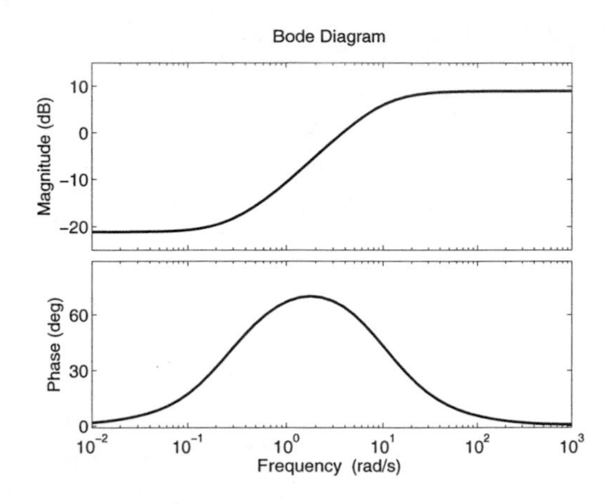

Figure 6.9. Bode plot of the compensator $C(s)$ (Equation (6.7)).

6.2.7 Compensator Splitting and Pilot Simulation

As anticipated (Section 6.7), a compensator replacing the human behavior is designed (Equation (6.7)) with the feedback force F_{OA} as input as depicted in Figure 6.2. Being the Equation (6.6) valid, the force F_{OA} and the aircraft distance from the corridor center line y_{OA} are linearly related. Thus the compensator $C(s)$ having a pole and a zero similarly to a proper Proportional Derivative Controller produces, roughly speaking, a control action that is proportional to distance form the center line and to its derivative. The human operator, for any regulation task of this kind, shows a proportional-derivative behavior in the sense that his/her command is proportional to the error (the distance from the center of the street) and to the derivative of the error (the center-line approach speed) as a kind of prediction of future error.

This allows to split the compensator into two actions: the haptic aid and the pilot action who acts thanks to the visual aid. Figure 6.10 depicts this concept.

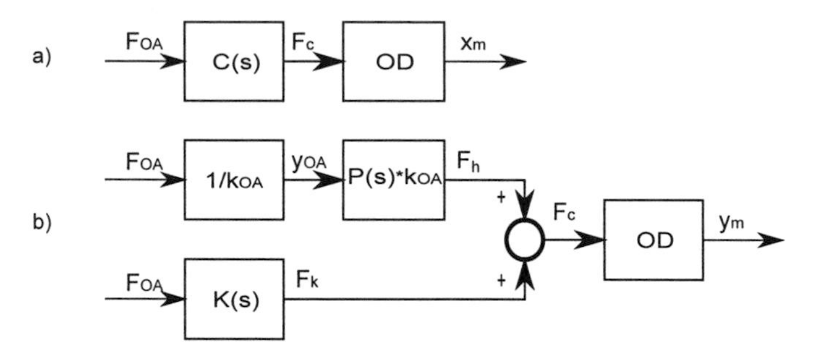

Figure 6.10. Compensator splitting.

The force F_h can be thought as the output of a pilot ($P(s)$ in Figure 6.10) that, summed up with the force F_k getting out from the latter part of the compensator ($K(s)$), gives the force F_c (actually the output of the compensator C_S). Given the linear relationship between F_{OA} and y_{OA}, the pilot's input becomes the distance from the center-line (y_{OA}) as he/she would receive from a visual feedback. Thus the pilot transfer function $P(s)$, which is designed to regulate the force F_{OA} to zero, has the same effect of regulating the distance from the center-line to zero. Thus the upper part of the Figure 6.10b can be thought as visual feedback, while the bottom part of Figure 6.10b can be thought as haptic feedback.

Note in the Figure 6.10 that $P(s)$ and $K(s)$ are designed in respect of Equation (6.8).

$$C(s) = P(s) + K(s) \tag{6.8}$$

Figure 6.11 re-depicts the baseline F-P scheme of Figure 6.2 with the employment of the Equation (6.8).

In order to define the values of the two components of the compensator in Equation (6.8), the possibility of providing a static haptic aiding system ($K(s) = const$) is evaluated first. In particular the first choice is heuristically

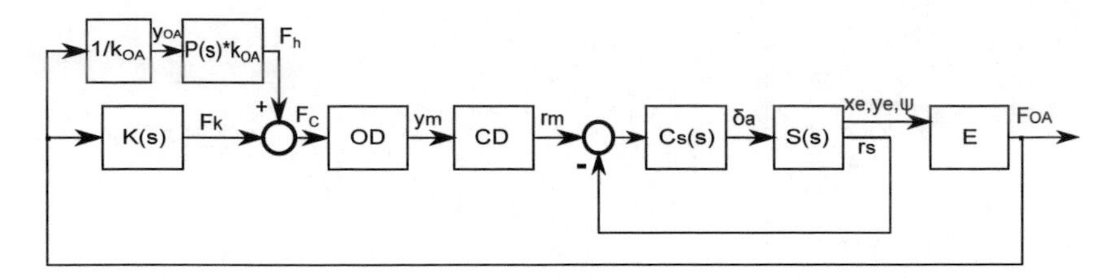

Figure 6.11. The baseline Force-Position scheme with compensator splitting.

set to $K(s) = 0.2$. $P(s)$ is easily found from Equation (6.8). A simulation of the scheme resulting from the splitting shows (see Figure 6.12) a big difference between F_h and F_k and, in particular this means that, in the first instants of the simulation the haptic component (F_k) is not so much relevant.

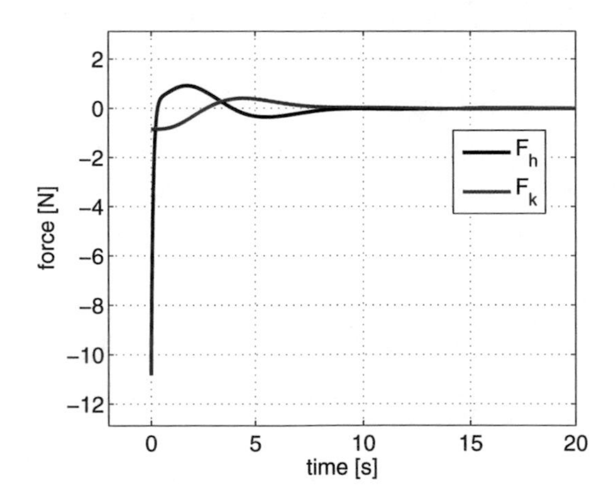

Figure 6.12. F_h and F_k time response when $K(s) = 0.2$.

Different choices for $K(s)$ are evaluated; Figure 6.13 shows the comparison between F_h and F_k for 3 different values of $K(s)$ (i.e. 0.1, 0.5, 0.9). $P(s)$ is once more calculated according to the Equation (6.8).

Note in Figure 6.13, F_h and F_k are opposite in sign for the most part of the simulation time, then it would not be desirable for the pilot

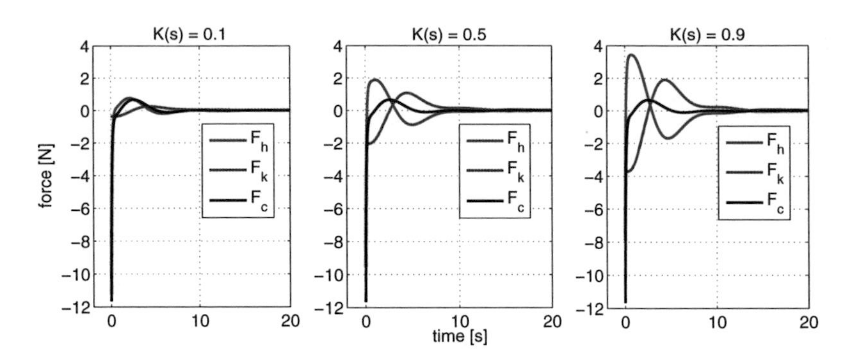

Figure 6.13. F_h and F_k time response when $K(s) = 0.1, 0.5, 0.9$ respectively.

to have the haptic force always in opposition (remember that DHA is the feedback philosophy employed in this Chapter). According to this, a relevant anticipatory effect or phase lead (as the derivative effect of standard industrial controllers) might be needed also in $K(s)$, otherwise the pilot would have to produce the whole anticipatory effect by him/herself. Then the choice of $K(s)$ as a percentage of $C(s)$ (see Equation (6.9)) is evaluated.

$$\begin{cases} K(s) = \gamma \cdot C(s) \\ P(s) = (1-\gamma) \cdot C(s) \end{cases} \tag{6.9}$$

Three sample values where tested: $\gamma = 0.25, 0.5, 0.75$ and the respective bode plots of the open loop system (pilot plus plant) of Figure 6.11 are depicted in Figure 6.14.

$\gamma = 0.5$ is selected being the slope of the pilot plus plant the closest to $-20dB/dec$ in observance of the crossover model [48] (see Section C.1). The feedback results to be divided exactly in two halves: one half the visual one, the other half the haptic one. Figure 6.15 shows the new time responses for F_h and F_k.

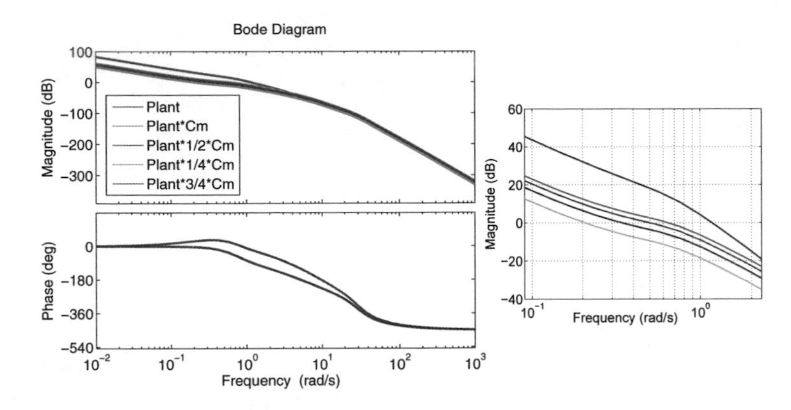

Figure 6.14. Bode plots of the open loop system of Figure 6.11. On the right a zoom in around the crossover frequencies to measure the slopes.

The final transfer functions chosen as $K(s)$ and $P(s)$ are shown in Equation (6.10).

$$K(s) = P(s) = \frac{1.4s + 0.4374}{s + 10} \tag{6.10}$$

6.2.8 F-P scheme: simulations

The capability of the designed haptic aiding force with respect to keeping the straight flight in the mentioned symmetric scenario (the long straight corridor between two buildings) is tested first. A simulation is run with the pilot out of the loop (i.e. the Omega Device end-effector moves by itself flying the aircraft into the corridor).

To initially perturb the state of the aircraft a non zero initial condition ($y_e = 5m$) is set for the simulation.

Figure 6.16a shows a sample simulation of Figure 6.2 scheme obtained using the identified transfer function of the Omega Device in place of the real device; the system shows a very fast and satisfactory response in the

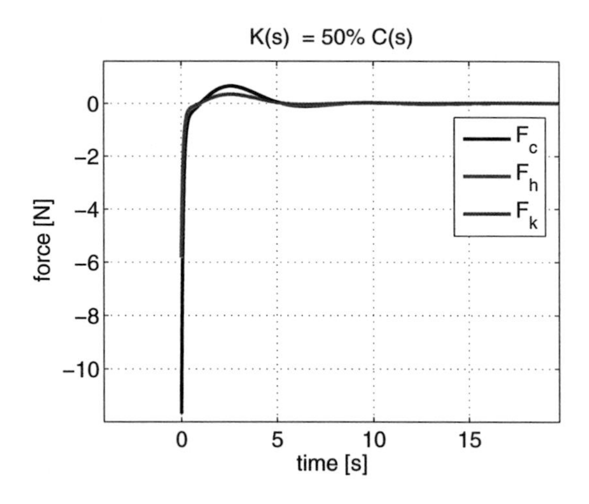

Figure 6.15. F_h and F_k time response when $K(s) = P(s) = 50\%C(s)$.

absence of delay and the *200 ms Delay* curve shows that the presence of the delay induces larger oscillations that anticipate instability with larger delays.

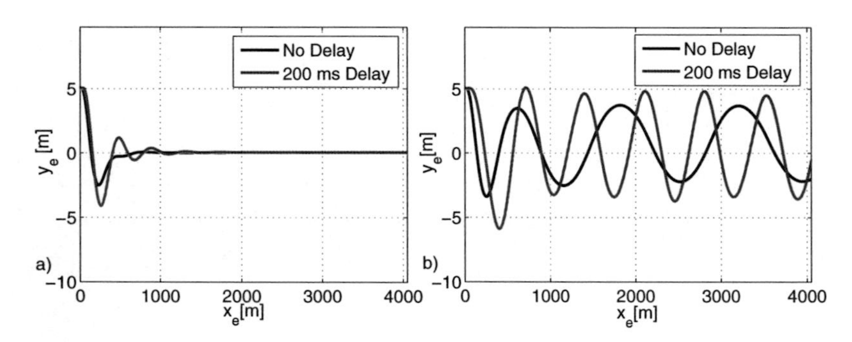

Figure 6.16. Path comparison (Figure 6.2 scheme) with and without time delay by using: a) the Omega Device model; b) the real Omega Device and the pilot out of the loop.

The same simulation is performed using the real Omega Device (without Pilot because his/her action is replaced completely with $C(s)$). Figure 6.16b shows an evident limit cycle that is due to the non linearities present in the real Omega Device and not captured by its linear identified model. The limit cycle is expected to vanish very likely when the pilot holds the stick because his/her arm provides additional inertia and damping. As a matter of fact, the same simulations run with the real Omega Device and the human operator in the loop in considered now. Figure 6.17a represents a simulation with and without time delay in which $F_{OA} = 0$ and the operator in the loop. These simulations show that the pilot without the haptic aiding does not produce a suitable trajectory (too much close to the obstacles).

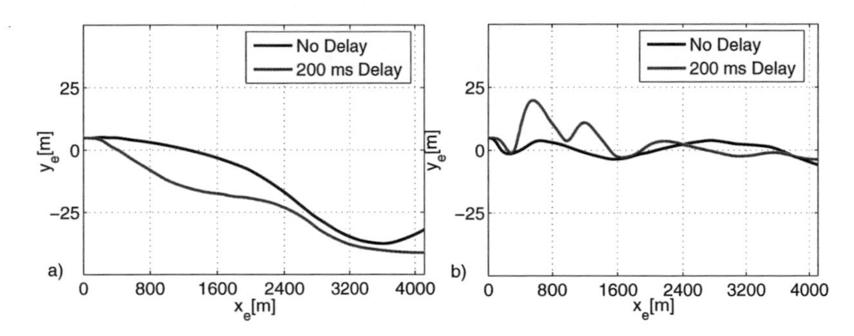

Figure 6.17. Path comparison (Figure 6.2 scheme) with and without time delay and the human operator in the loop. a) $F_{OA} = 0$; b) $F_{OA} \neq 0$.

Conversely Figure 6.17b represents a simulation with and without time delay, $F_{OA} \neq 0$ and the operator in the loop (in this case the output F_h of the block $P(s)$ in the scheme 6.10b is disconnected).

By comparing the Figure 6.17a with the Figure 6.17b, note how important is for the human operator the presence of the haptic feedback F_k which helps him/her to stay in the middle of the corridor. Clearly the presence of delay makes the task harder and produces more oscillations around the condition where the haptic force is zero (the middle of the street).

6.3 The Wave Variables Approach

Often stability problems induced by delays are tackled in teleoperation systems using wave variables.

The typical Force-Position scheme with wave variables is shown in Figure 6.18.

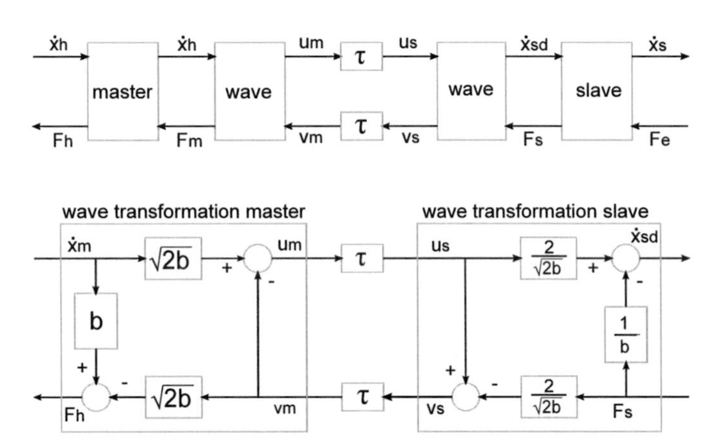

Figure 6.18. The typical wave variable scheme [8].

The wave variables are calculated starting from the power variables, velocity and force, through the equations depicted in the blocks "wave transformation master/slave" of Figure [8] where τ is the communication delay; the subscripts "h", "e", "m" and "s" represent respectively the human operator, the environment, the master and the slave variables. The wave variables technique is based on the concepts of energy and passivity. Intuitively, a system is passive if it absorbs more energy than it produces. In fact, the power in the communication link is defined through the difference between the power input (velocity and force from the master side) and the power output (velocity and force from the slave side). If a system is passive than it is stable. The delays in the communication link may destroy the stability of the system by producing energy; in fact, the communication delays shift the signals and the product between the just mentioned power variables may change and may bring to the production of energy in the communication link. The wave variables were shown to produce always a positive energy (the input energy is bigger than the output energy); thus, passivity and stability are theoretically ensured. For details on this technique refer to [8, 52, 63].

In this work, a preliminary evaluation of the effect of the wave variables transformation is performed.

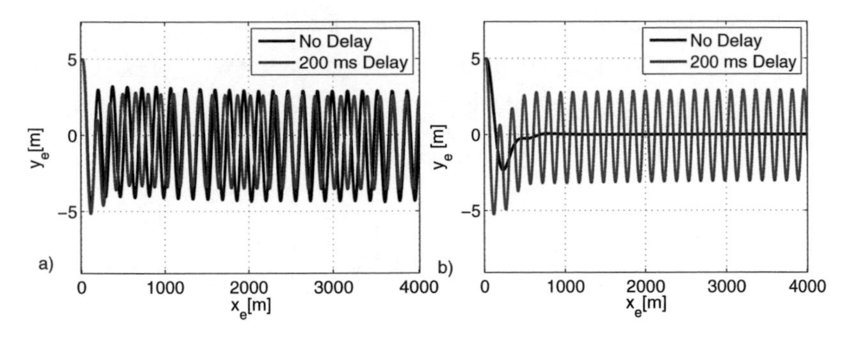

Figure 6.19. The wave variable simulations with and without time delay and the operator out of the loop by using: the real Omega Device (a); the Omega Device transfer function (b).

Figure 6.19a shows a simulation with and without time delay with the operator out of the loop and the real Omega Device.

Figure 6.19b shows the path comparison with and without time delay of the scheme obtained by employing the Omega Device transfer function (i.e. a simulated haptic device) in place of the real Omega Device.

It is evident that the use of wave variables does not add a significant improvement in the present implementation. In fact, in the present application an important hypothesis about the employment of the wave variables is not satisfied: the slave is not passive. As a matter of fact, the aircraft (the slave) has an unstable dynamics. Then, a different solution should be employed to reduce the instability effect of the delay.

6.4 Fa-P scheme

In order to mitigate the effect of the delay over the aircraft trajectory, an admittance-based teleoperation scheme [56] is setup. As said, the admittance control architecture is generally employed with admittance-type devices. The haptic device employed in this work is (see Section A.2) an impedance-type device characterized by very light-weight construction with low inertia and friction. However, high dynamic properties (i.e. high stiffness), usually characterizing the admittance-type devices, were set for it in order to simulate a FBW stick-type. This is the main reason for the implementation of an admittance-based teleoperation scheme. The compensator splitting described in Section 6.2.7 is employed and Figure 6.20 shows the exploited Fa-P admittance scheme.

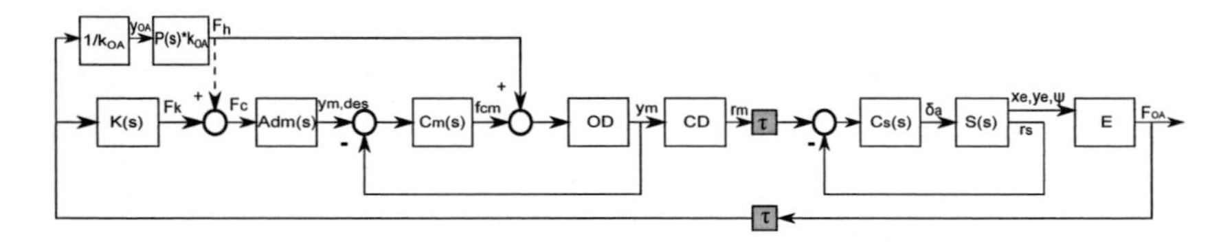

Figure 6.20. The admittance scheme Fa-P.

Positions are sent from master to slave and forces from slave to master. Admittance-type controller is used to control the master device, while the slave is controlled by a position controller.

The force F_h can be thought as the output of a pilot ($P(s)$ in Figure 6.20) that is summed up with the force f_{mc} getting out from the local master compensator ($C_m(s)$) in one side and with F_k getting out from the latter part of the compensator ($K(s)$) in the other side. The pilot transfer function $P(s)$ acts in a way that makes the force F_{OA} to be zero (i.e. in the middle of the street where $y_{OA} = 0$). The force F_c is fed to an admittance block, $Adm(s)$, to produce a reference signal for the master side, $y_{m,des}$, to help the human operator in the obstacle avoidance task (see Section 6.4.1 for details). Clearly, the F_h signal can be feed-forwarded to the admittance block only during the simulation (i.e. using the Pilot mathematical model $P(s)$) as long as the real haptic device would need a force sensor not actually available on the stick. The rest is the same as in Section 6.2.

6.4.1 Admittance and local master controller

Equation (6.11) shows the admittance transfer function employed in the $Adm(s)$ block of Figure 6.20.

Commonly admittance-type devices are controlled by using a so called position-based admittance. Such an architecture is used hereby with an impedance-type device having, as said, a target dynamics (high dynamic properties, i.e. high stiffness) usually characterizing the admittance-type devices: the haptic device aim in this work is, in fact, to simulate a FBW stick-type and thus to render a target dynamics (e.g. desired mass, stiffness and damping behavior) of a FBW stick-type. This is achieved by implementing admittances in the form of simple mass-spring-damper systems:

$$y_{m,des}(s) = \frac{F_k(s)}{M_d s^2 + B_d s + K_d} \tag{6.11}$$

In the Equation (6.11) the values of the desired mass M_d, desired damping B_d and desired stiffness K_d are chosen in order to obtain suitable stability properties of the system with the operator out of the loop (see later).

During free space motion only the master impedance given by M_d, B_d and K_d is active. In "contact" with the environment (when the distance between the aircraft and the obstacles is less than r_e) the stiffness k_{OA} is felt and both master and slave controllers influence the impression of the remote environment. M_d, B_d and K_d have to be selected to guarantee stability of the overall system. On this account, in order to reduce the number of control parameters, M_d was set to the actual mass of the end-effector ($0.1Kg$). k_{OA} was carefully selected through experimental trials (simple simulations run with different k_{OA} values) and by interrogating the subjects about the perception of the environment. k_{OA} is set to $0.167N/m$, a good tradeoff between having tangible forces when the aircraft is flying around the middle of the corridor and not even too strong forces when the aircraft is flying close to the obstacles' walls. The maximum value of the obstacle avoidance haptic force nearby the obstacles' walls is about $8N$, as said, in order to have some margin (about $4N$) for the action of the haptic aid exerted by the master compensator (see Section 6.2.6). B_d and K_d were selected via stability considerations (see later).

Equation (6.12) shows the local master controller transfer function employed.

$$C_m(s) = \frac{37.56s + 981.1}{s} \tag{6.12}$$

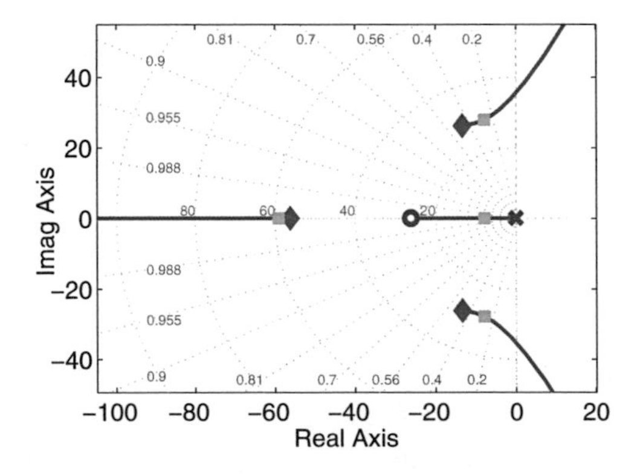

Figure 6.21. The master root locus to design the compensator $C_m(s)$.

It was designed in the linear domain using the Evans' Root Locus tool in order to have an appropriate response time (about $0.6s$). Figure 6.21 shows the root locus used for the design. See depicted the open loop poles as blue diamonds, the compensator roots (poles as crosses, zeros as circles) in red and the closed loop poles as green squares. In order to design the compensator $C_m(s)$ of Equation (6.12), the identified model of the Omega Device (Equation (B.2)) was employed.

Admittance target dymanics tuning

The system of differential equations (Equation (6.13)) describes the dynamics of Figure 6.20 and the Equations (6.12), (B.2) and (6.11) are employed.

$$\begin{cases} f_{cm}(t) = 981.1 \cdot \int (y_{m,des}(t) - y_m(t))dt + 37.56 \cdot (y_{m,des}(t) - y_m(t)) \\ f_{cm}(t) = \frac{1}{7.118} \cdot \ddot{y}_m(t) + \frac{26.7}{7.118} \cdot \dot{y}_m(t) + \frac{864.18}{7.118} \cdot y_m(t) \\ F_k(t) = M_d \cdot \ddot{y}_{m,des}(t) + B_d \cdot \dot{y}_{m,des}(t) + K_d \cdot y_{m,des}(t) \end{cases} \tag{6.13}$$

Stability and transient response analyses are carried out in order to find out suitable values as B_d and K_d. Operating the Laplace transform on the Equation (6.13), a linear state-space model is extracted (see Equation (6.14)).

$$\begin{cases} s\mathbf{X}(s) = \mathbf{A} \cdot \mathbf{X}(s) + \mathbf{B} \cdot \mathbf{U}(s) \\ \mathbf{Y}(s) = \mathbf{C} \cdot \mathbf{X}(s) + \mathbf{D} \cdot \mathbf{U}(s) \end{cases} \tag{6.14}$$

$\mathbf{X(s)}, \mathbf{U(s)}$ and $\mathbf{Y(s)}$ are respectively the Laplace transform of the state vector $\mathbf{x(t)}$, the input $\mathbf{u(t)}$ and the output $\mathbf{y(t)}$ vectors; $\mathbf{A} = f(B_d, K_d)$,

$\mathbf{B} = f(M_d)$, while $\mathbf{C}$ and $\mathbf{D}$ are respectively an identity matrix and a null matrix. The state vector is shown in Equation (6.15), while the only input is the force F_k.

$$\mathbf{x}(t) = \begin{bmatrix} y_{m,des}(t) \\ \dot{y}_{m,des}(t) \\ y_m(t) \\ \dot{y}_m(t) \\ \ddot{y}_m(t) \end{bmatrix} \tag{6.15}$$

The asymptotic stability of the overall system is analyzed: the linear time invariant system described by the state space model of Equation (6.14) is asymptotic stable if, for all eigenvalues λ_i of the system matrix $\mathbf{A}$, holds:

$$Re\{\lambda_i(\mathbf{A})\} = \sigma_i < 0, \forall i. \tag{6.16}$$

B_d and K_d were chosen in order to have negative eigenvalues' real part of matrix $\mathbf{A}$ (Equation (6.16)) and good free motion master sensations (see Figure 6.22).

Note in Figure 6.22 that three of the five system poles are not effected by the parameters. As concerning instead the two poles affected by variations in B_d and K_d notice that when $B_d = 0.2$ or $B_d = 5$ the master side dynamics is excessively oscillatory or slow respectively. When $K_d = 50$ or $K_d = 400$ it is respectively too looser or too stiff resulting to be unpleasant to the pilots. The values of $B_d = 1Kg/s$ and $K_d = 200Kg/s^2$ (wide size cyan poles in Figure 6.22) were selected, a good tradeoff between good stability properties and free motion master sensations.

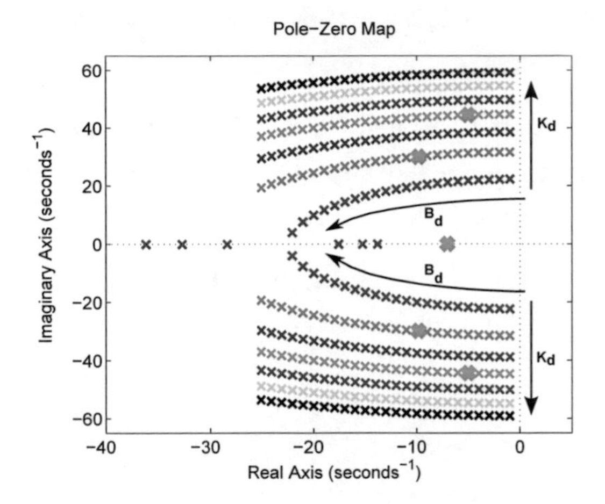

Figure 6.22. Pole map of the system for $K_d = [0.2 : 0.2 : 5]$ and $B_d = [50 : 50 : 400]$. In wide size cyan the $B_d = 1Kg/s$ and $K_d = 200Kg/s^2$ poles.

6.4.2 Fa-P scheme: simulations

Figure 6.23a shows the aircraft trajectory between the buildings in a simulation with and without time delay when the dotted line F_h to the admittance block is employed (see Figure 6.20).

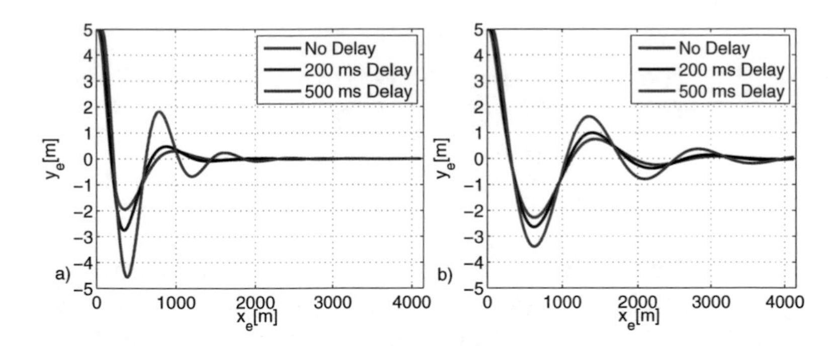

Figure 6.23. Admittance scheme (Figure 6.20) simulations with and without time delay when the dotted line is: a) employed; b) cut.

Figure 6.23b shows the aircraft trajectory between the buildings in a simulation with and without time delay when the dotted line F_h to the admittance block is cut (see Figure 6.20).

By comparing Figure 6.23a and Figure 6.23b, note that summing up F_h to F_k, that is having a force sensor on the stick, provides better transient properties to the system.

In Figure 6.24 notice a simulation with the human operator in the loop (then $P(s) = 0$ in Figure 6.20) with and without time delay.

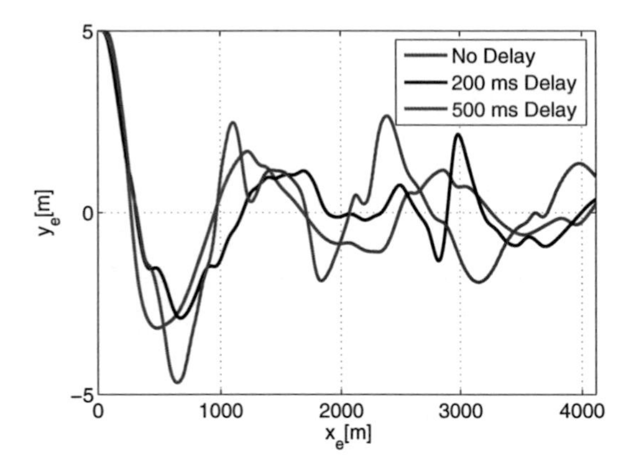

Figure 6.24. Admittance scheme (Figure 6.20) simulations with and without time delay with the real Omega Device and the human operator in the loop.

By comparing the Figure 6.24 with the Figure 6.23a and the Figure 6.23b, it might be better to sum up the force of the human operator to the haptic feedback F_k as in Figure 6.20. Unfortunately the Omega Device employed for the experiments does not have a force sensor, thus an observer (see Subsection 6.4.3) for the human force is designed in order to implement something similar to the scheme of Figure 6.20.

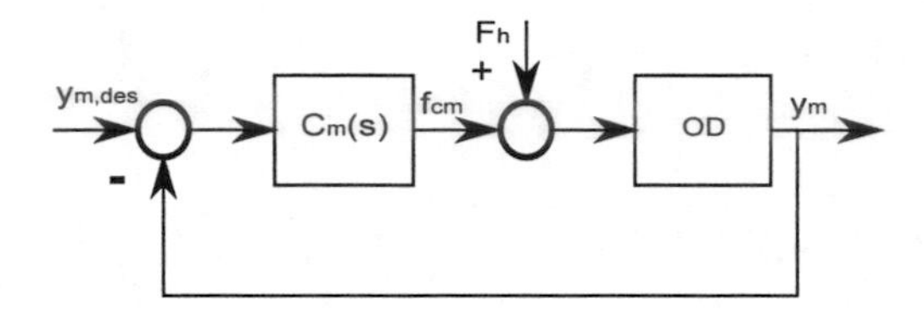

Figure 6.25. Scheme employed to build up the human force observer.

6.4.3 The human force observer

Figure 6.25 shows the inner part of the Master control loop. The human force acts as unknown input, the system OD_i (the identified model of the Omega Device, OD) is known with a certain approximation, the system $C_m(s)$ (the master admittance controller) is known exactly, the signal $y_{m,des}$ is internally generated and then known exactly, and the signal y_m is measured by the haptic device sensors, then it is known with approximations. Equation (6.17) shows the transfer function from $y_{m,des}$ to y_m in Figure 6.25.

$$y_m(s) = \frac{C_m(s) \cdot OD_i(s)}{1 + C_m(s) \cdot OD_i(s)} \cdot y_{m,des} + \frac{OD_i(s)}{1 + C_m(s) \cdot OD_i(s)} \cdot F_h \qquad (6.17)$$

Solving for F_h, it is possible to define the final expression of the observer transfer function, $O(s)$ as in Equation (6.18):

$$O(s) = \hat{F}_h = \frac{1 + C_m(s) \cdot OD_i(s)}{OD_i(s)} \cdot y_m - C_m(s) \cdot y_{m,des} \qquad (6.18)$$

where $\hat{F}_h$ is the observed F_h.

Figure 6.26 shows the scheme employed for the implementation of the observer (Equation (6.18)).

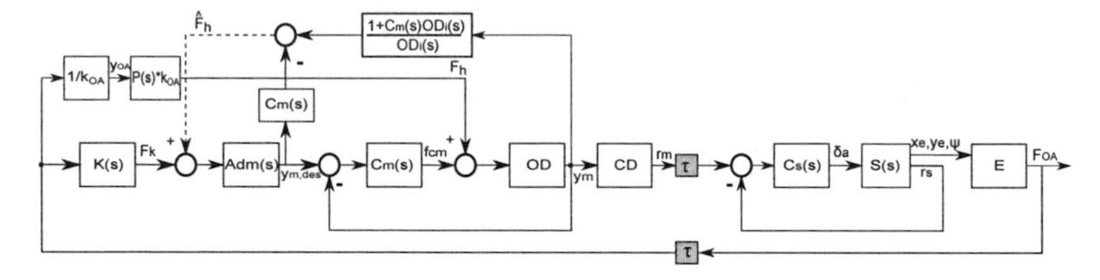

Figure 6.26. The observer scheme.

Figure 6.27 shows the observer scheme in which also the visual feedback (delayed by τ seconds) is shown explicitly.

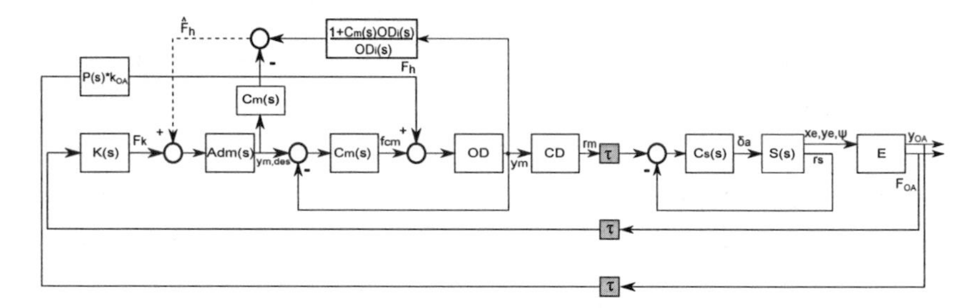

Figure 6.27. The observer scheme with visual feedback.

Since the first component of Equation (6.18) is an improper transfer function, through the addition of two high frequency poles it is made proper in order to be able to implement it (see the Bode plot in Figure 6.29). Figure 6.28 shows the comparison between F_h and $\hat{F}_h$ during a sample simulation.

In order to implement the observer with the real Omega Device (which provides the signal y_m as a discrete signal) in the loop a discretized model of the observer dynamics is needed. The Tustin approximation, preferred for filter approximations, is employed. Figure 6.29 shows both the effect of making the observer transfer function proper and the quality of the discrete approximation.

In order to evaluate the observer performance, since no force sensor is

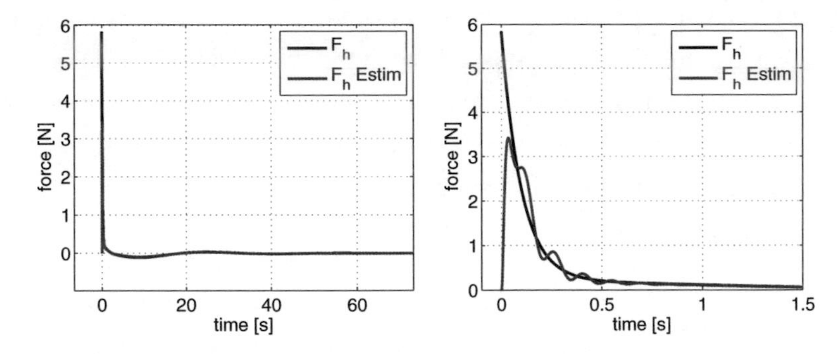

Figure 6.28. Observer validation (Figure 6.26 scheme) by employing the Omega Device model. Comparison between F_h and $\hat{F}_h$. On the right, the zoom in on the origin.

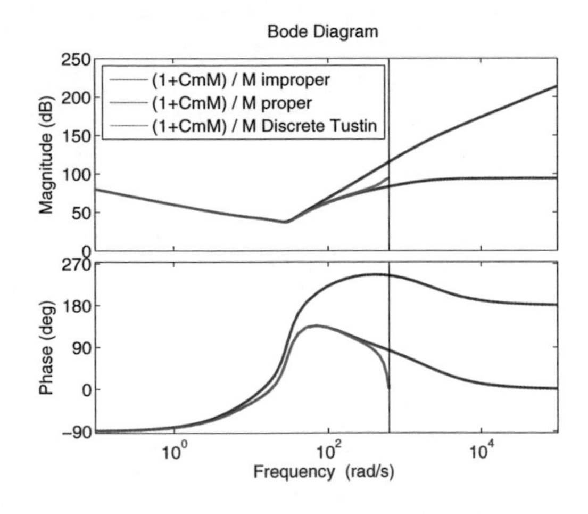

Figure 6.29. Bode plot comparison of the first term of Equation (6.18). In red, blue and green respectively the improper, the proper and the discrete transfer functions.

available to compare with, two simulated human operator force scenarios are defined (see Figure 6.30): in the first one F_h is set to be a constant force ($2N$ magnitude); in the second one the observer is asked to estimate a sinusoidal force which magnitude (about $2N$) and frequency (about $25s$) are similar to the oscillating forces produced during a simulation with the aircraft. The time responses are depicted in Figure 6.30. The red line is obtained with the identified model of the Omega Device, the magenta line is obtained with the real Omega Device in the loop. Notice that in Figure 6.30a the observer produces a signal which mean value is highly similar to F_h. Then, the observer works adequately although some spike is present; these are caused by the noisy signal y_m, the sensor output displacement of the real Omega Device.

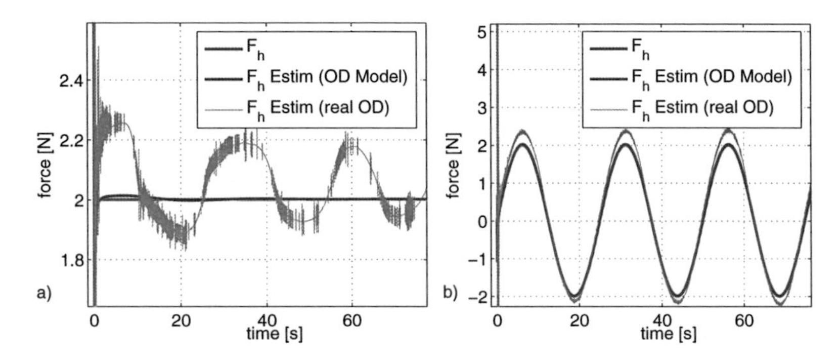

Figure 6.30. Observer validation (Figure 6.26 scheme) by employing both the Omega Device model and the real one. Comparison between F_h and $\hat{F}_h$. Zoom around the origin. In the legend OD is for Omega Device. Instead of the human operator a forcing function is employed: a) $2N$ constant force; b) $2N$ amplitude and 25 seconds period sinusoidal force.

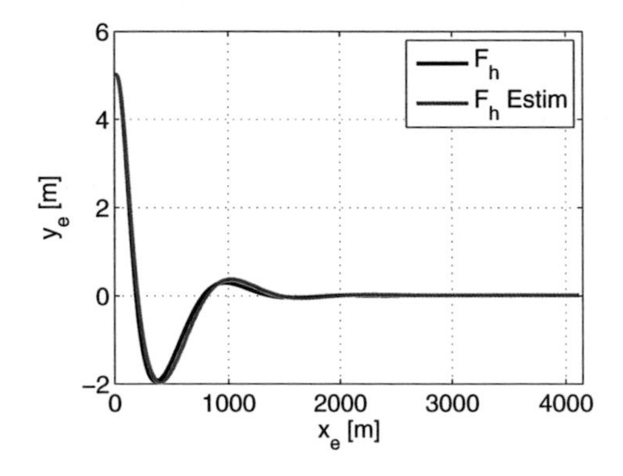

Figure 6.31. Simulations comparison (Figure 6.26 scheme) by using F_h and $\hat{F}_h$.

Figure 6.31 shows two simulations where the system output (lateral position of the aircraft) is compared when using the real F_h and the observed $\hat{F}_h$; it appears that the results achieved in both cases are comparable.

Figure 6.32a shows three simulations obtained using the observer (scheme of Figure 6.26). Figure 6.32b compares the results obtained by running the simulation of Figure 6.26 with and without the dotted line $\hat{F}_h$, and with a time delay of 500 ms. It appears clearly, also by direct comparison with figures 6.23a and 6.23b, which present the same simulations achieved with the exact knowledge of the human force, that the addition of the observer has a beneficial effect in terms of transient response of the system.

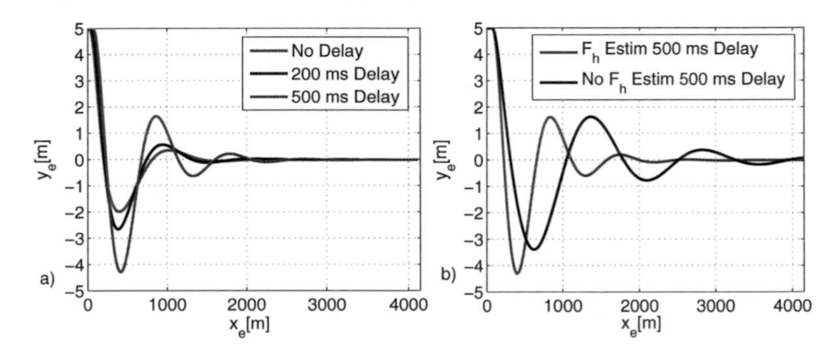

Figure 6.32. Observer scheme (Figure 6.26) simulations by employing the Omega Device model: a) the dotted line ($F_h Estim$) is employed (0,200ms,500ms delay); b) 500 ms delay simulations comparison with and without the dotted line.

Figure 6.33 shows the improved system stability under 500 ms delay with the employment of the admittance controller and the observer with respect to the baseline scheme (FP teleoperation). The same Figure compares the simulation outputs using both the real F_h and the observed $\hat{F}_h$; it is visible that the observer works properly and that the degradation of the transient performance when using the observer is minimal.

Finally, Figure 6.34 exhibits three trials with the human operator in the

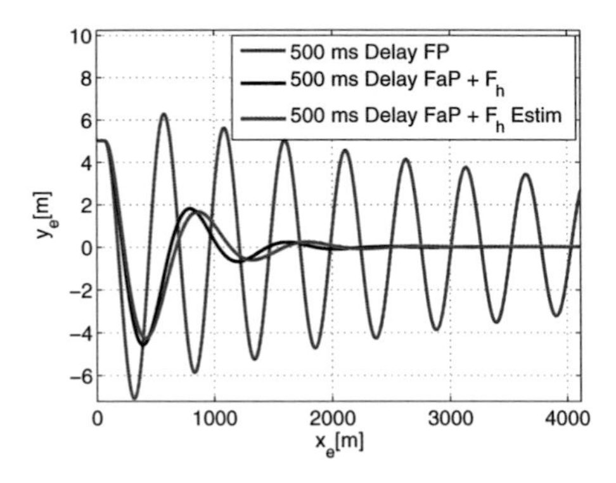

Figure 6.33. FP and FaP (Figures 6.2 and 6.26 schemes) simulations comparison under 500 ms delay by employing the Omega Device model.

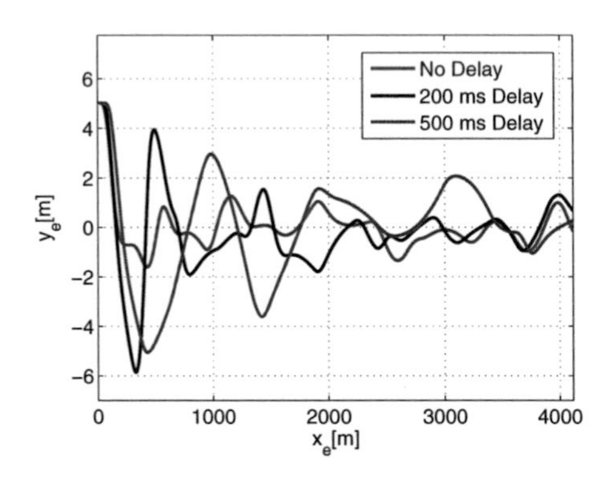

Figure 6.34. Admittance scheme (Figure 6.27) simulations with and without time delay with the human operator in the loop.

loop. By direct comparison between figures 6.34 and 6.24, even though a throughout analysis with a relevant number of trials and test pilots would be needed, it arises that transient performance improves with the adoption of the observer, and that the transient performance achievable with the FaP admittance scheme outperforms those of the FP scheme.

In order to evaluate the performance of the system over a more complex environment, four trials are run within the obstacle environment designed in Chapter 4. The simulations are performed using the scheme described in Figure 6.27 (dotted line included) with the real Omega Device and the pilot in the loop. Figure 6.35 shows two trials in which $F_{OA} = 0$ (i.e. no Haptic aiding). Figure 6.36 shows two trials in which $F_{OA} \neq 0$ (i.e. the haptic aiding is active).

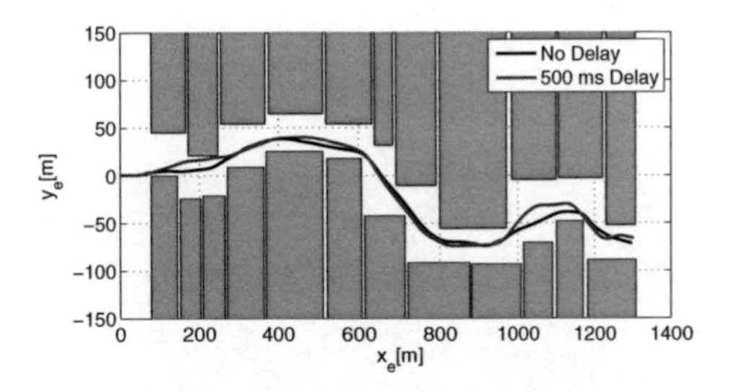

Figure 6.35. Simulations (Figure 6.27 scheme) with pilot in the loop and $F_{OA} = 0$.

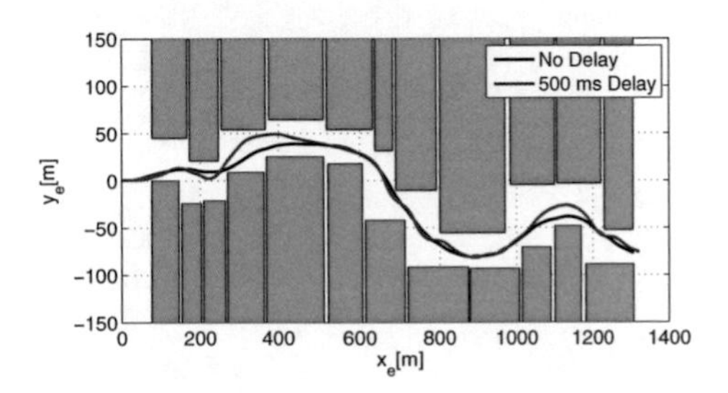

Figure 6.36. Simulations (Figure 6.27 scheme) with pilot in the loop and $F_{OA} \neq 0$.

By comparing Figure 6.35 with Figure 6.36 notice that there are no significant differences in pilot performance (i.e. number of collisions).

Then, in order to make the task more difficult, some fog in the visual display is added; the resulting visibility became thus extremely low and the pilot, de facto, had to rely much more on the haptic cues. Figure 6.37 shows two trials in which $F_{OA} = 0$. Figure 6.38 shows two trials in which $F_{OA} \neq 0$.

By comparing Figure 6.37 with Figure 6.38 remark that, at least for the

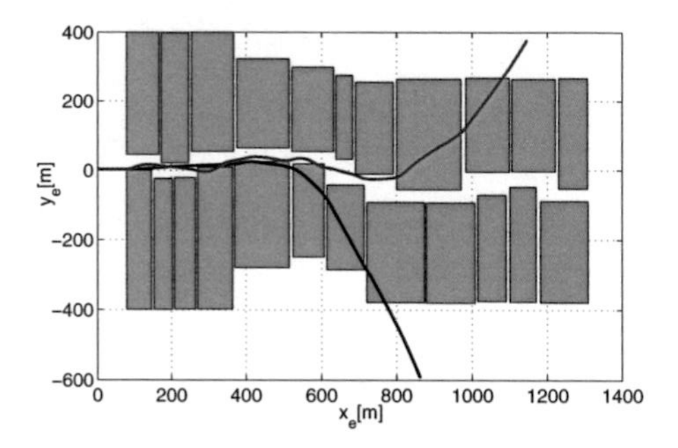

Figure 6.37. Simulations (Figure 6.27 scheme) with pilot in the loop and $F_{OA} = 0$ in fog conditions. The black line shows the *No Delay* trial. The red line shows the *500 ms Delay* trial.

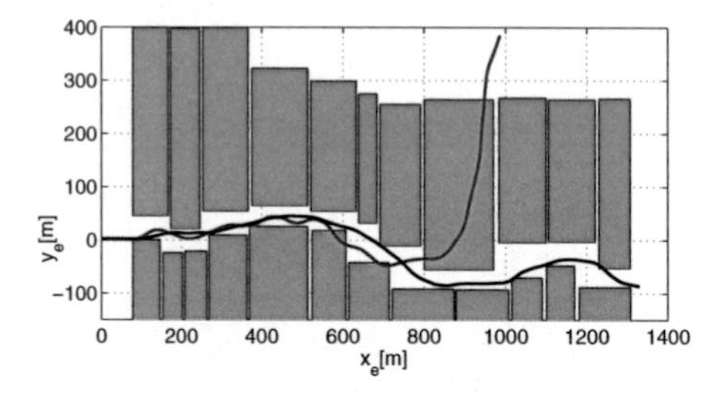

Figure 6.38. Simulations (Figure 6.27 scheme) with pilot in the loop and $F_{OA} \neq 0$ in fog conditions. The black line shows the *No Delay* trial. The red line shows the *500 ms Delay* trial.

No Delay trajectory, the haptic feedback is very important in improving the pilot performance. The *500 ms Delay* trajectory in Figure 6.38 appears slightly better than the corresponding without haptic feedback, but suggests at the same time that an improvement in the haptic feedback is probably needed.

7 Conclusions

Both remote piloted systems for Unmanned Aerial Vehicles and Fly-By-Wire systems for manned aircraft (which the present study could be applied as well, see Section 3.1) do not transfer to the pilot important information or cues regarding the state of the aircraft and the loads which are being imposed by the pilot's control actions. These cues have been shown to be highly responsible for pilot situational awareness.

Thus, the opportunity of artificially reintroducing them in the pilot control device arose and brought to the necessity of designing an artificial feel in the control device [25].

Moreover, the bandwidths of modern flight control systems approach the pilot's own sensing and actuation systems and this could bring to undesirable effects like pilot-induced-oscillations (PIO).

It has been shown in the past [17] that, since Situational Awareness is created through the perception of the situation (SA First Level), the quality of SA is extremely dependent on how the person directs attention and how attention to information is prioritized based on its perceived importance.

Due to all the reasons above, it is essential to increase the knowledge of human-machine challenges among system developers and users [53].

Furthermore, blaming crashes and mishaps on human error is usual in UAV teleoperation field and the wrong assumption that humans cause most errors brings many people to believe that errors can be avoided by removing the human and by employing full automation [12]. On the contrary, several UAVs incidents and crashes have been attributed to automation errors or loss of situational awareness because the human has been "automated out of the loop" (Human System Interface deficiencies) [53].

There has been little research on UAV "cockpit" design and its impact on the human operator. A lot of research is still required in evaluating different designs of UAV interfaces that optimize operator performance abilities. Human and automation teamwork, when efficient, could achieve levels of performance and safety beyond that of the human or automatic systems alone. Automation entities are not flexible as humans are. The high rates of mishaps and crashes happening today in the UAVs field would have been significantly lower if human-machine teamwork would have been given more attention in the design evolution of UAVs control laws [53].

The automation should be designed in a way that better support human performance, reduce the workload and support the decision making processes. Thus, investing in a human machine interface design tailored on the human needs would improve the operator situational awareness and maybe the performance as well.

All the previous considerations suggest in particular that the force-feel system design is still an important issue now that the performance capabilities of modern aircrafts have increased exponentially and these are the reasons for which **the force feel are now to be considered as**

part of the vehicle!

Thus a question arises: which are the specifications and the behavior of the "ideal" artificial force-feel system?

The maximum forces an individual can exert is an example of how important is to tailor the artificial feel directly on the human being.

Due to the previous consideration, it appears that taking into account the human operator natural behavior in the design of new generation aiding system might be a winning point. In the present work, to better address the haptic aid design a review and a classification of the haptic aids present in the scientific literature was made first. Two haptic aids classes were defined and were given the name of Direct and Indirect Haptic Aid. Afterwards, the idea to consider the human operator natural behavior in the haptic feedback design, was made through the introduction of the Indirect Haptic Aid for disturbance rejection and/or obstacle avoidance tasks. Thus, an artificial feel system, that drew its inspiration in the mechanical force-feel systems for fixed-wing aircraft in which important informations are felt by the pilot through the control device, was employed and developed.

Although haptic feedback is used in various areas (included UAV teleoperation) and with different goals, seems that application of haptic feedback in UAV teleoperation for both collision avoidance and path following in low airspace in the presence of external disturbances such as wind gusts has been not investigated so far. The haptic information should not only map the environmental constraints or location goals but also the external wind conditions because the gusts (vertical or lateral) in presence of obstacles could be extremely dangerous for the structural safety of the UAV. Thus, the haptic feedback should be needed for mapping both natural and environmental constrains.

Particularly, when the visual information is hinder or limited (e.g.

obstacles outside the field of view or foggy weather conditions), the haptic feedback might compensate for the lack of visual information also in the presence of external disturbances as wind gusts.

As a matter of fact, when the UAV is approaching an obstacle in the presence of fog, for example, a sudden maneuver is needed in order to avoid the collision. In the presence of fog, in fact, the distance at which the obstacle is seen is shorter than the same distance in case of good visibility conditions and the effective time for avoiding the obstacle is considerably reduced. By employing the haptic canal of information in addition to the visual canal, the remote pilot would "see" the obstacle approaching earlier through the haptic feedback than through the visual one.

This would increase the Situational Awareness and the safety of teleoperation.

All of the Indirect Haptic Aids introduced in this work (Conventional Aircraft Artificial Feel, Obstacle Avoidance Feel and Mixed-CAAF/OAF) are an attempt of readily inform the remote pilot about the presence of a potential danger which could bring to the mission failure.

In fact, the main goal of the IHA-based approaches developed herein was to improve the situational awareness about the state of the drone hopefully showing that a performance improvement would also come as a consequence. As a matter of fact, CAAF would inform the pilot about an external disturbance affecting the UAV; OAF would inform the pilot about the environmental constraints and Mixed-CAAF/OAF would inform the pilot about both environmental constraints and external disturbances affecting the UAV.

Furthermore, the present work exhibited an improvement of performance when IHA-based approach was employed:

- the Variable Stiffness CAAF was tested and shown to increase the performance with respect to the absence of haptic feedback at all;

- Force Injection CAAF was proved to increase the performance in terms of instinctive response to a stimulus in pilots without any previous training on the experiment with respect to the conventional haptic aids.
- OAF and Mixed-CAAF/OAF were exhibit to increase the performance in terms of collisions avoidance with and without the presence of wind gusts with respect to the conventional haptic aids.

Such performance improvements were compared to those available with the other commonly used, and published in the scientific literature, approaches which fall in the DHA category.

The goal of the DHA simulators employed in this work was not to obtain state-of-the-art performance, but to serve only as a comparison term for the IHA simulators.

During the implementation of the DHA simulators for comparison with the IHA approach, it was found out that the design of a DHA based augmentation scheme is highly task dependent.

In the CAAF VS DHA Experiment, for example, a reference altitude had to be chosen for DHA and a compensator capable of holding it was designed. The compensator gain was then reduced in order to leave some authority to the pilot: the aim of this work was aiding teleoperation and not designing an automatic control system. Reducing the gain of the DHA-compensator would make the pilot useful.

In the OAF VS DHA Experiment an attempt to design the DHA compensator to be a slightly more task-independent was made. As a matter of fact, the pilot was given a certain freedom in choosing the path. In this experiment, what made the performance difference between the

IHA and the DHA concepts was probably that DHA forced the operators to fly at a distance (from the obstacles) in which the force field was not excessively strong and, for this reason, more comfortable; with the IHA force the pilot was instead free to fly pretty close to the obstacles because no force helping to avoid them was available.

In the MIXED CAAF/OAF VS DHA Experiment, the DHA was designed in order to make the aircraft lateral acceleration null; this behavior would efficiently reject the lateral wind gust as a stand-alone compensator but it was proved to be unsafe in terms of number of collision. Furthermore, this approach would fail or, at least, would exhibit an undesirable behavior in the case the pilot's intention is to perform a maneuver that creates a lateral acceleration like, for example, a sideslip maneuver.

All of the previous considerations make the DHA-approach rather reasonably to be an "almost automatic system" having roughly the same drawbacks as autonomous systems: DHA design is highly task-dependent and it would likely remove the pilot from the decision making process. The IHA-based approach, instead, would leave to the pilot full authority in the decision making process and, being deeply important that the pilot runs and at least supervises the whole mission, it would keep higher his/her attention on the task and, as a consequence, all the UAVs mishaps causes would hopefully be reduced and an improved safety would be reached. The IHA-based approach would leave space to the pilot in case his/her intention is not known but depends only on last moment decisions reached through some unknown cognitive process.

It might appear singular to compare two Haptic Aiding schemes producing force sensations which have opposite sign on the same task. In fact, the experiments conducted so far showed that the participants (both professional pilots and naïve subjects) can control the aircraft within both DHA and IHA approaches without a-priori instructions or training but the IHA-based ones produced better results. IHA systems appeared to be more intuitive to be handled.

In general, human responses to external stimuli are highly conditioned by the required processing operations. In line with this, some motor responses are more 'automatic' (less affected by cognitive factors) and occur with shorter latency. For instance, saccades are more 'natural' than antisaccades [38]. The stretch reflex, which is a reflex contraction of a muscle in response to passive longitudinal stretching, is a highly automatic motor response that is believed to be the spinal reflex with the shortest latency [73]. Application of the IHA concept to both the disturbance rejection and the obstacle avoidance problems, subject of this dissertation, produced a force stimulus to which the operator must, in general, oppose.

Several other examples could be built following the IHA concept and would lead to similar results: a stimulus to be counterbalanced and overtaken. Thus, the IHA concept, which requires a reaction in opposition to stimuli rather than compliance, might therefore be more 'natural' for the system because it very likely exploits the highly automatic and fast stretch response [10, 83]. The preliminary analyses of the psychophysical implications of this research suggest that the type of motion task required by the IHA concept could be thought like being composed by a stretch reflex in response to initial force peak (caused by the gust and/or obstacle edge), together with a higher-level response caused by the experience in rejecting wind gust disturbances and by the visual cues. Would this be true, the author could conclude that, at least for certain types of applications, an Haptic feedback which operates accordingly to the IHA concept (i.e which produces stimuli to be opposed) would result more natural to be understood and followed by the operator, and possibly would provide better task performance, than a similar system built following the DHA concept.

The teleoperation target of this research was also tested in the presence of time delay in the communication link. The employed setup resulted in a "non-classical" teleoperation scheme, since the feedback was related to the distance from obstacles and not to the force that results from the interaction with the environment. This is the reason for which the results obtained in scientific literature (usually applied to classical teleoperation architectures) [56, 57, 8, 52, 63], needed an adaptation to be ported to this application.

Since the knowledge of the force the pilot exerts is needed in the control law implementation and no force sensors were available in the actual control device, an observer was designed and proved capable to estimate it (at least simulated human force injected into the actual haptic device in software). The resulting admittance scheme plus the human force observer was shown to be able to provide good transient performance both in simulations and with the human operator in the loop.

A

Experiments Setup

The present work was mainly conducted at the Max Planck Institute for Biological Cybernetics of Tübingen under the sponsorship of the director Professor Heinrich H. Bülthoff and of the University of Pisa, with the supervision of Professor Lorenzo Pollini.

All the experiments were conducted in a dark room for making the participants to focus their attention on the experiments.

A computer with a 24 inch liquid crystal display (LCD) screen and a control device, namely the Omega Device, in one side (left or right according to the subjects preference) played the role of a fixed-base flight simulator.

The layout of the experiment environment is shown in Figure A.1.

An LCD screen was employed to display the simulation scenarios: an EFIS display (see Figure 3.3) for the disturbance rejection experiments and a synthetic view of a street with buildings (see Figure 4.1) for the obstacle avoidance experiments.

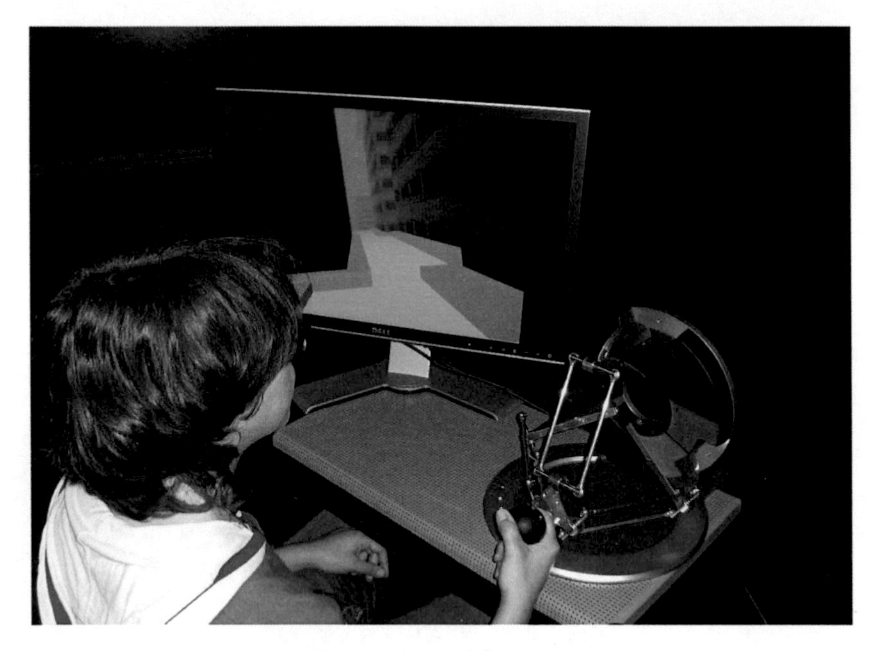

Figure A.1. The experimental setup.

During the experiments, the subjects were seated in the darkened room and the experiment coordinator was next to them following the trials.

The experiments were run on MatLab and the Simulink Toolbox which operated at 200 Hz. The visualization ran at 20 Hz (see Section A.3 for details); the haptic loop around 3000 Hz (see Section A.2 for details).

A.1 The Aircraft Model

In all the experiments the Flight Dynamics and Control Toolbox [68] was employed. It is a graphical software environment for the design and analysis of aircraft dynamics and control systems based upon the complex non linear model developed in Simulink from M.O. Rauw.

It is distributed exclusively across the Internet through the website *http://www.dutchroll.com*. In particular, for the disturbance rejection experiments the full non linear model was employed. While a linearization around a trim condition was used for the obstacle avoidance experiments.

The Beaver De Havilland Canada DHC-2 was the simulated aircraft employed. It is a fixed-wing aircraft with single propeller engine. The fully non linear model consisted of 12 Ordinary Differential Equations (ODE).

In the disturbance rejection experiments the aircraft model was trimmed to fly horizontally and the trim conditions were:

$$\begin{cases} V \simeq 50m/s \\ H = 300ft \\ \gamma = 0deg \Rightarrow \alpha = \theta \end{cases} \tag{A.1}$$

The engine ran at constant $1800rpm$. To obtain the trim conditions shown in Equation (A.1), the elevator had to be deflected $\delta_{e,trim}$ and the manifold pressure (the thrust in case of the aircraft under consideration) was kept constant to MP_{trim}:

$$\begin{cases} \delta_{e,trim} = -0.2565deg \\ MP_{trim} = 25"Hg \end{cases} \tag{A.2}$$

In all the simulations, the thrust was kept constant at the value of Equation (A.2), while the elevator was deflected around the trim value of the same equation.

In the disturbance rejection experiments, the only considered dynamics was the longitudinal one and the lateral input (the aileron deflection δ_a and the rudder deflection δ_r) were kept at zero.

In the obstacle avoidance experiments, only the lateral dynamics was considered. Thus, δ_a and δ_r were employed for controlling the aircraft lateral dynamics. New trim values were needed:

$$\begin{cases} \delta_{e,trim} = 0.5856deg \\ \delta_{a,trim} = 0.0661deg \\ \delta_{r,trim} = -2.1933deg \end{cases} \tag{A.3}$$

A.1.1 Technical Data

Some aircraft technical data are shown below. The *flight envelope* depicts the boundaries of aircraft loading and flight conditions within which operation of the aircraft is satisfactory and beyond which some aspect becomes unacceptable. It depicts at some particular velocities the maximum load factor that could be introduced by the maneuvers remaining handling qualities, engine behavior and structural loads acceptable. Figure A.2 depicts an example of the simplest flight envelope.

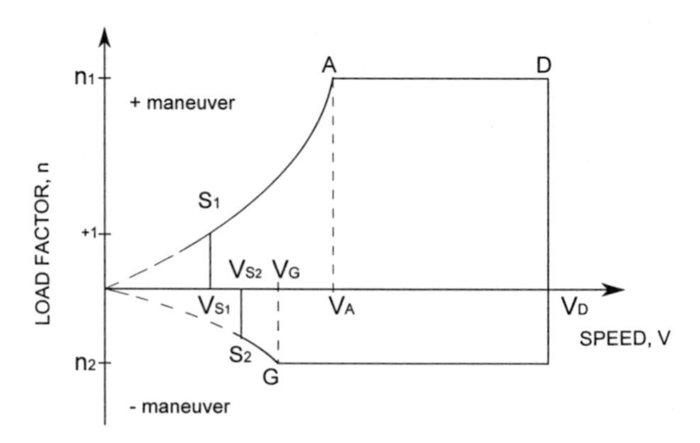

Figure A.2. The flight envelope.

V_{S1} and V_{S2} are the stall speeds in upright and inverted flight respectively. They can be interpreted as the minimum speed at which the steady horizontal flight ($n = 1$ in right side up flight and $n = -1$ in inverted flight) is possible. Such velocities correspond to the maximum $C_{L,max}$ and minimum $C_{L,min}$ (unknown) lift coefficients and are calculated through:

$$\begin{cases} n_1 \cdot W = \frac{1}{2}\rho V_{S1}^2 S C_{L,max} \\ n_2 \cdot W = \frac{1}{2}\rho V_{S2}^2 S C_{L,min} \\ W = m \cdot g \end{cases} \tag{A.4}$$

being:

$$\begin{cases} C_{L,max} = 2.2 \\ S = 23.23m^2 \\ \rho = 1.225Kg/m^3 \\ m = 2288.231Kg \\ g = 9.81m/s^2 \\ n_1 = 3.8 \\ n_2 = -1.52 \end{cases} \tag{A.5}$$

S is the wing area, ρ the air density, m the aircraft mass, g the gravity acceleration, n_1 and n_2 are respectively the maximum positive and negative (inverted flight) load factors.

V_A and V_G are the design maneuvering speeds in right side up and inverted flight respectively.

V_{S1} and V_{S2} are calculated through $n = 1$ and $n = -1$ respectively; V_A and V_G, velocities corresponding to the maximum and minimum lift coefficients for the maximum and minimum aircraft load factors, are calculated for $n = n_1$ and $n = n_2$ respectively.

V_D is the design diving speed. Being this value unknown, it was hypothesized to be:

$$V_D = V_{NE} + 10\% V_{NE} \tag{A.6}$$

where V_{NE} ($80.25m/s$) is the never exceed speed.

A.1.2 Aicraft Natural Modes

By linearizing the complex non linear model around the trim conditions of Equation (A.1), the following longitudinal transfer function is obtained:

$$H_{Lon} = \frac{5.985s^3 + 2.2364s^2 - 1298.2s - 27.607}{s^5 + 5.7716s^4 + 15.229s^3 + 0.73354s^2 + 0.74834s} \tag{A.7}$$

Figure A.3 shows the Bode plot of the transfer function of Equation (A.7).

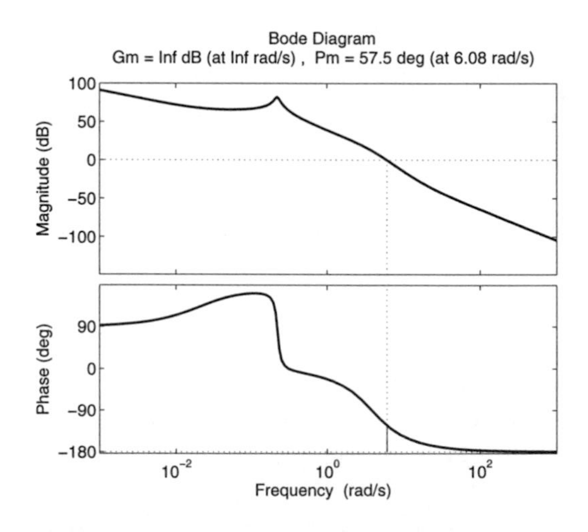

Figure A.3. Bode plot of the Beaver longitudinal dynamics.

While Figure A.4 shows the pole-zero map of Equation (A.7).

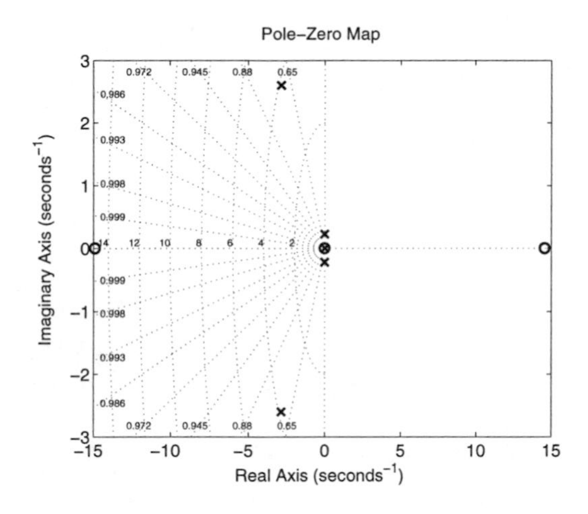

Figure A.4. Pole-zero map of the Beaver longitudinal dynamics.

The low damped couple of complex and conjugate poles (ζ about 0.06) represent the Phugoid mode poles. While the well damped one (ζ about 0.75) represent the Short Period mode poles. The Beaver longitudinal dynamics contains a non minimum phase zero as well.

By linearizing the complex non linear model around the trim conditions of Equation (A.3), the following lateral transfer function is obtained:

$$H_{Lat} = \frac{-9.9877s^3 - 10.4132s^2 - 6.1385s + 0.0184}{s^4 + 8.1578s^3 + 10.2490s^2 + 11.8186s + 0.6961} \tag{A.8}$$

Figure A.5 shows the Bode plot of the transfer function of Equation (A.8).

While Figure A.6 shows the pole-zero map of Equation (A.8).

The couple of complex and conjugate poles (ζ about 0.5) represent the

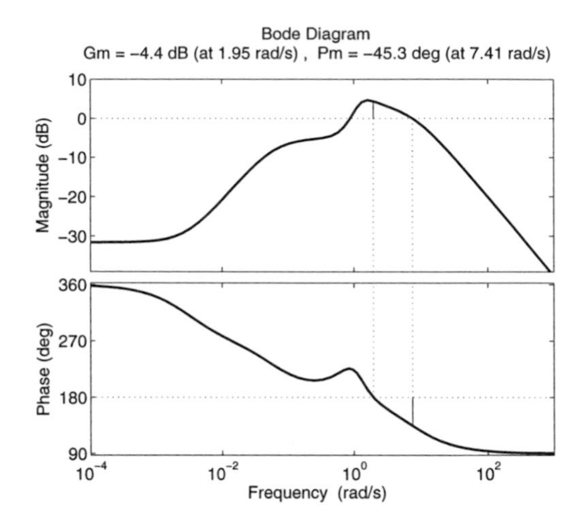

Figure A.5. Bode plot of the Beaver lateral dynamics.

Dutch Roll mode poles. While the real poles represents the low frequency (long time constant) Spiral mode and high frequency (fast time constant) Roll Subsidence mode.

The pilot or the autopilot has to damp the above longitudinal and lateral aircraft natural modes.

A.2 The Haptic Device

The control device employed in this work is the Omega Device (omega.3 produced by the Force Dimension, Switzerland): a high precision force feedback device classified as an **impedance-like** haptic device. Some technical data is shown in Table A.1.

An S-Function was built to make the Omega Device communicate with the PC. The control loop which implemented the haptic device dynamics was realized in software. A Microsoft windows application constructed a

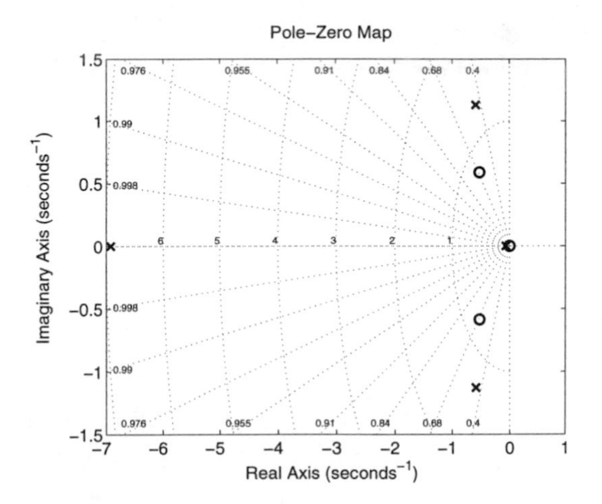

Figure A.6. Pole-zero map of the Beaver lateral dynamics.

workspace	translation	⊘ 160 x 110 mm
forces	translation	12.0 N
resolution	translation	$<$ 0.01 mm
stiffness	closed-loop	14.5 N/mm

Table A.1. The Omega Device Specifications.

thread which implemented all the haptic algorithms and was set to run as fast as possible; the software was executed in a dual core machine, thus one of the two cores was essentially devoted to executing the haptic control loop. A statistical measure of the thread execution frequency was recorded: the haptic loop execution frequency resulted to be around 3000 Hz. Due to this high activation frequency, no clitches or other undesired disturbances were noticed in the rendered force.

The stiffness and damping for respectively the longitudinal (x-axis of Figure 3.2) and the lateral (y-axis) constants were heuristically chosen and are shown in the Table A.2:

Longitudinal Stiffness	$K_{S,x} = 240$ N/m
Longitudinal Damping	$K_{D,x} = 6Ns/m$
Lateral Stiffness	$K_{S,y} = 120$ N/m
Lateral Damping	$K_{D,y} = 6Ns/m$

Table A.2. The implemented stick features.

A.3 The 3D Visualization System

The out of window view of Figure 4.1 and the EFIS display of Figure 3.3, were produced using DynaWORLDS. DynaWORLDS is a software project born at the Department of Energy and Systems Engineering at the University of Pisa, from an idea of Lorenzo Pollini and Gaetano Mendola, and later developed by DynamiTechs (*www.dynamitechs.com*) to build a low-cost, comprehensive distributed simulation and Synthetic Environment (SE) visualization toolset.

Mathworks Real Time Workshop ©can be used effectively to automatically generate C code to be used for simulation. Network connections are based on TCP/IP and UDP/IP protocols, but the same data stream could be sent on any transmission channel simply by coding appropriate device drivers.

Synthetic environments can be created using an integrated framework of scene design, object animation, and control panel design. The world, or scene, can be designed by means of 3-D objects, whose geometry and surface properties are imported by commercial CAD file formats, lights, and cameras. Each object can be connected to a motion channel that affects its position, orientation, and scaling in 3-D space; can be linked to any one other so as to inherit some of its features (a robotic arm); and its position can be tracked with a trail. A link can be established even among objects, cameras, and lights so that one object can bring cameras (inside vision from a vehicle) and lights (car's headlight). Motion channels are the animation sources for the scene; a channel is the abstraction of a data stream that may have several sources: files, network sockets, I/O boards, or input devices such as joysticks or buttons. With motion channels, all these sources can be mixed to obtain very complex object animation.

A control panel can then be designed interactively on-screen using output devices: camera views, various instruments such as pointers, light indicators, or artificial horizons, and so on.

DynaWORLDS is also capable of drawing nonfixed geometry objects; trails, smoke, fog, clouds, or typical augmented reality tools such as a guidance tube or data superimposed on recognized objects on the screen can be drawn using appropriate graphical plug-ins. Furthermore, particular transformations such as nonlinear scaling, squeezing (useful for displaying collisions between elastic objects), or bending (vital for representation of flexible structures) are only possible with custom software.

Figure A.7. Snapshot of a F-22 aircraft simulator.

One of the most important requirements of a hardware-in-the-loop simulation environment is its capability to incorporate various input and output devices to allow full integration of hardware components and

Figure A.8. Snapshot of an underwater vehicle simulator.

software-simulated systems. Only with custom software device drivers and the adoption of a common communication standard it is possible to virtually connect heterogenous systems in their interfaces and sampling time. Every new real-world device can be put in the simulation loop with ease and without relying on the nonstandard interfaces adopted by other commercial applications. In the end, complete control over the final rendering makes environmental features such as viewing through fog or turbid water, or even the reproduction of night vision device images, feasible. Figures A.7 and A.8 show a couple of simulation examples.

B
Omega Device Identification

This Appendix presents the results of the model identification procedure applied to the Omega Device.

The longitudinal transfer function $OD_{i,x}(s)$ of the actual Haptic device used in the disturbance rejection experiments (see Chapter 3) is shown in Equation (B.1). It was identified by using frequency sweeps (from 0.0262 to 10 Hz) and the Empirical Transfer Function Estimate (ETFE) technique [46].

$$OD_{i,x} = \frac{3}{s^2 + 8.413s + 902.7} \tag{B.1}$$

The stiffness and damping constants (for both longitudinal and lateral identification procedure) are shown in Table A.2.

Figure B.1 depicts in blue the real Omega Device Bode plot and in red the identified model Bode plot.

An example of the time response comparison between the real and the

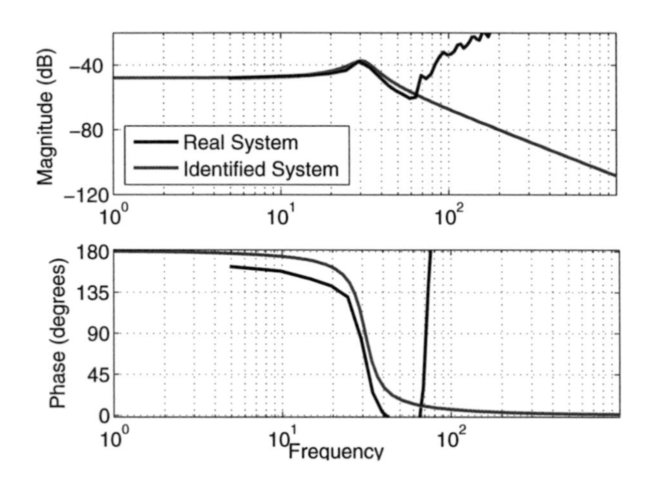

Figure B.1. Real vs Identified Omega Device longitudinal Bode plots.

identified Omega Device for the longitudinal case obtained for a frequency sweep with amplitude of $3.2N$ is shown in Figure B.2.

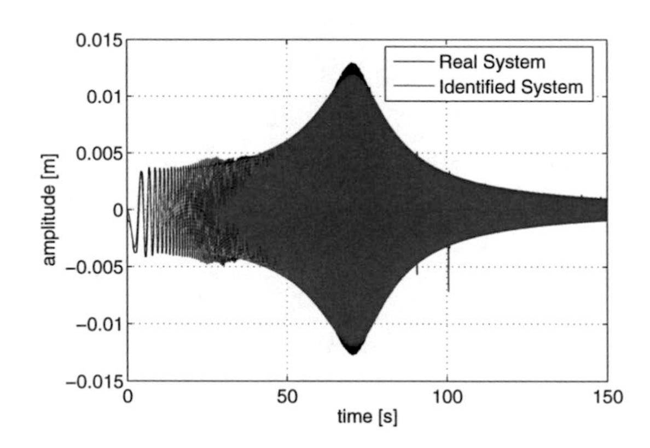

Figure B.2. Time responses comparison between Real and Identified Omega Device longitudinal dynamics.

As concerning the Omega Device lateral dynamics identification, the same procedure as above was employed. It resulted in the transfer function $OD_{i,y}(s)$ of Equation (B.2) and it was employed in the obstacle avoidance experiments (see Chapters 4, 5 and 6).

$$OD_{i,y} = \frac{7.118}{s^2 + 26.76s + 864.8} \tag{B.2}$$

C DHA Compensators Design

This Appendix argues the design of the DHA compensators employed in this work.

In particular, the Section C.1 shows the design of the DHA disturbance rejection compensator for the longitudinal dynamics employed in Section 3.5.2; the Section C.2 shows the design of the DHA disturbance rejection compensator for the lateral dynamics employed in Section 5.2.

C.1 DHA for Longitudinal Disturbance Rejection

McRuer presented a detailed study of human operator dynamics in compensatory tasks [48]. This research concentrated upon the effects of forcing function bandwidth and controlled element dynamics upon human operator describing functions (transfer functions and remnant). One very important product of the reported research is the "crossover model" of the human operator or pilot. This model essentially describes the ability of the human to adapt to different controlled elements and random appearing command inputs with different bandwidths.

It is mainly based on the assumption that in the area of the whole system crossover the human will adjust to different plant dynamics to yield the same human plus plant dynamics that is a simple integrator behavior. The Hess structural model which focus on the ability to adapt to different vehicle dynamics [28] is based on the McRuer crossover model.

The plant in this case is a combination of the control device dynamics and of the aircraft dynamics to control.

As concerning the longitudinal dynamics, the longitudinal model of the control device of Equation (B.1) and the linearized aicraft longitudinal model of Equation (A.7) are considered.

The pilot has to control the plant longitudinal dynamics (Equation (C.1)) represented by the series of the previous transfer functions:

$$H_{Lon} = \\ = \frac{-17.96s^3 - 6.709s^2 + 3895s + 82.82}{s^7 + 14.18s^6 + 966.5s^5 + 5339s^4 + 1.375e004s^3 + 668.5s^2 + 675.5s} \tag{C.1}$$

The plant Bode plot is depicted in Figure C.1.

The slope of the plant Bode plot (Figure C.1) around the crossover frequency (about $0.5rad/s$) is close to $-40dB/dec$. Thus the pilot model has to produce around the same frequency a positive slope of about $20dB/dec$ in order to get a simple integrator behavior (i.e. a $-20dB/dec$ slope) of human plus plant dynamics.

Hess in [28] gives detailed indication on how to calculate the value of

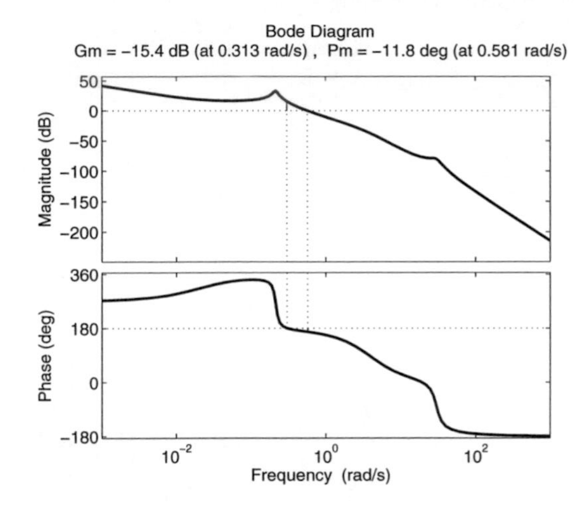

Figure C.1. The plant Bode plot with stability margins.

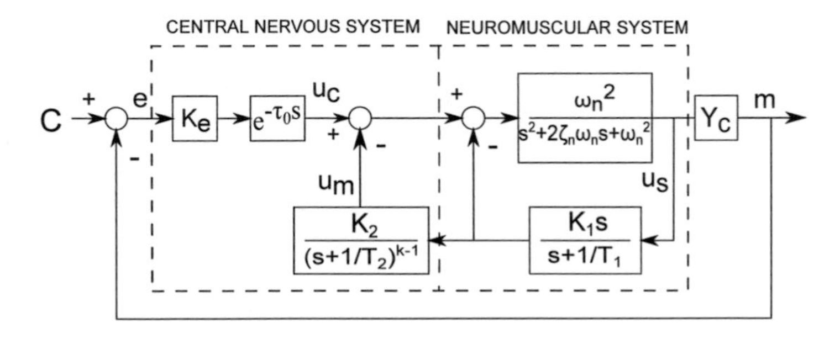

Figure C.2. The Hess Structural Model [28].

the new the human plus plant crossover frequency (in this case $3.18 rad/s$) in case of $1/s^2$ (current case) behavior of the plant. Through [28] is possible to calculate all the constants needed to build the human model (Figure C.2) that results in Equation (C.2):

$$C_{Lon}(s) = \frac{6452s^2 + 2584s}{s^4 + 14.75s^3 + 209.5s^2 + 1089s + 13.04} \tag{C.2}$$

The longitudinal plant copensated and not Bode plots are depicted in

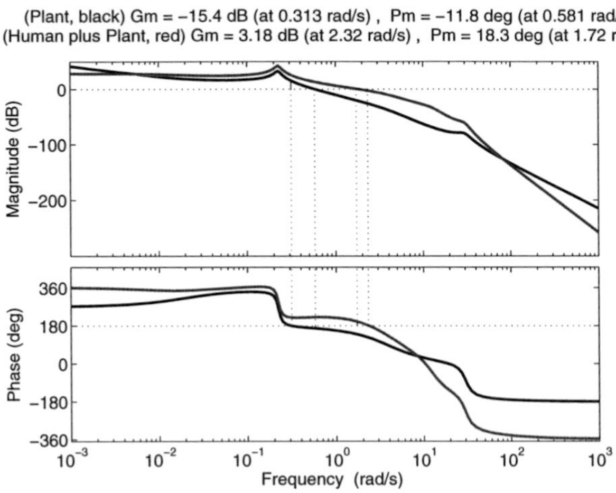
(Plant, black) Gm = −15.4 dB (at 0.313 rad/s) , Pm = −11.8 deg (at 0.581 rad/s)
(Human plus Plant, red) Gm = 3.18 dB (at 2.32 rad/s) , Pm = 18.3 deg (at 1.72 rad/s)
Magnitude (dB)
0
−100
−200
Phase (deg)
360
180
0
−180
−360
10^-3
10^-2
10^-1
10^0
10^1
10^2
10^3
Frequency (rad/s)

Figure C.3. Longitudinal Plant vs Human plus Plant Bode plots.

Figure C.3.

C.2 DHA for Lateral Disturbance Rejection

In this case a simpler compensator (a *phase lead network*) was employed.

The lateral plant to control (Equation (C.3)) is constituted by the series of the lateral control device identified dynamics of Equation (B.2) (the input is the output force of the compensator and the output is the control device lateral displacement which represents the aileron deflection) and the linearized ($sin(angle) \simeq angle$ and $cos(angle) \simeq 1$) lateral aircraft dynamics represented in Figure 5.2 (the input is the aileron deflection and the output is the lateral acceleration $\ddot{y}_B$) by considering $v_W = 0$.

$$H_{Lat}(s) = \\ = \frac{-697.4s^3 - 727.1s^2 - 428.6s + 1.284}{s^8 + 134.9s^7 + 4585s^6 + 1.167e005s^5 + 7.433e005s^4 + \\ +9.283e005s^3 + 1.025e006s^2 + 6.02e004s} \tag{C.3}$$

The compensator $C_{Lat}(s)$ (Equation (C.4)) was designed in the linear domain using the Evans' Root Locus tool in order to have a good response time (about $0.6s$).

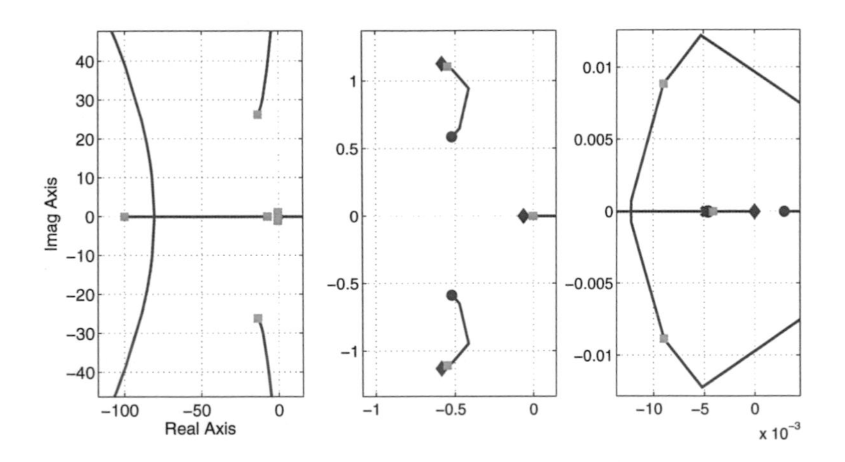

Figure C.4. The Evans' Root Locus used to design the compensator $C_{Lat}(s)$. From the left, the second and the third figures are a zoom in on the origin.

$$C_{Lat}(s) = \frac{102.1s + 0.4717}{s + 0.0048} \tag{C.4}$$

Figure C.4 shows the root locus used for the design. See depicted the open loop poles as blue diamonds, the open loop zeros as blue circles, the compensator roots (poles as crosses, zeros as circles) in red and the

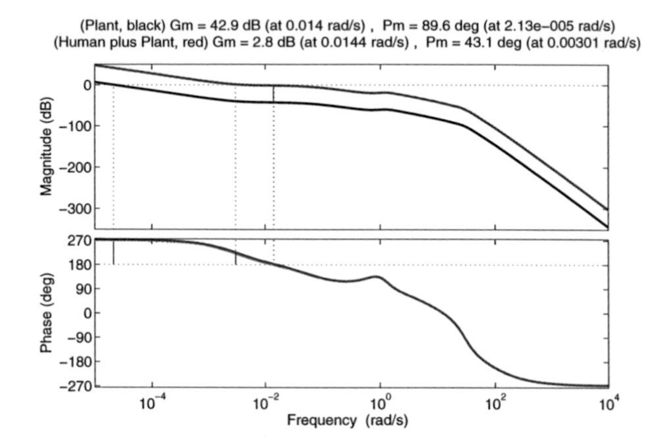

Figure C.5. Lateral Plant vs Human plus Plant Bode plots.

closed loop poles as green squares.

The lateral plant compensated and not Bode plots are depicted in Figure C.5.

Experiments Background

D.1 The CAAF Experiment

Eighteen naïve subjects participated in the experiment.

As said in Section 3.6.1, in the CAAF experiment, object of this section, a simple control task was prepared: the aircraft is initially flying leveled in trimmed condition at constant altitude ($300ft$); an artificial impulse (simulating a vertical wind gust) on the elevator induces the aircraft to initiate a motion according to its Phugoid mode.

The experiment consisted of three different force conditions: *No Force* condition with only compensation of gravity activated on the end-effector, *Simple Force* with Variable Stiffness CAAF (Equation (3.19)) and *Double Force* (force module twice as much as in the Simple Force condition). Each condition was run as a separate block, i.e., the experiment consisted of three successive blocks.

The trials' (24 of 120 seconds each, 8 trials per force condition) order of presentation of the blocks was counter-balanced according to the Latin

Squared Method (see Table D.1).

In total, the experiment lasted from 60 to 90 minutes (including instructions and breaks between blocks).

Subj No.	Block 1	Block 2	Block 3
1	1	2	3
2	1	3	2
3	2	1	3
4	2	3	1
5	3	1	2
6	3	2	1
7	1	2	3
8	1	3	2
9	2	1	3
10	2	3	1
11	3	1	2
12	3	2	1
13	1	2	3
14	1	3	2
15	2	1	3
16	2	3	1
17	3	1	2
18	3	2	1

Table D.1. The blocks order of presentation for each of the 18 participants. 1: *NoF*; 2: *Single VS CAAF Force*; 3: *Double VS CAAF Force*.

No instructions were given about the force conditions to test the natural reaction of the subjects to the three different conditions.

Before starting the real experiment each participant had to run a 5 minutes trial about the first block condition.

D.1.1 Instructions to subjects

You are going to pilot a simulated aircraft through the use of the Omega Device which is a force feedback device, i.e. when you move its end-effector you can feel a force feedback. During the experiment you will watch at the electronic instrument display: on the right side you see the altitude, in the center the artificial horizon in which the angle between the aircraft and the horizon is shown (when this angle is zero it means that you're flying in the horizontal plane). The only dynamics you have to control is the longitudinal dynamics (to make the aircraft to go up or down). To do this you need to move the stick forward or backward only: you have to pull the end effector to climb (to go up), to push the end-effector to dive (to go down); lateral or vertical end-effector movements do not affect the aircraft trajectory. The first 10 seconds of each trial, the aircraft is flying at constant altitude (300 ft). At time $9.5s$ a $0.5s$ duration disturbance (a vertical wind gust) affects the aircraft. The task of the experiment is to bring the aircraft at the initial altitude condition and to keep it there as much as possible.

D.1.2 Subjects detailed results

In Figure D.1, the three types of force were grouped: blue for No Force condition, green for VS CAAF-Simple Force condition, red for VS CAAF-Double Force condition.

The correspondence with the results provided in Section 3.6.1 is evident.

D.2 The CAAF VS DHA Experiment

Seven professional pilots participated in the experiment.

As said in Section 3.6.3, in the CAAF VS DHA experiment, object of this section, a simple control task was prepared: the aircraft is initially flying leveled in trimmed condition at constant altitude ($300ft$); three severe vertical wind gusts, which induce the aircraft to initiate a motion

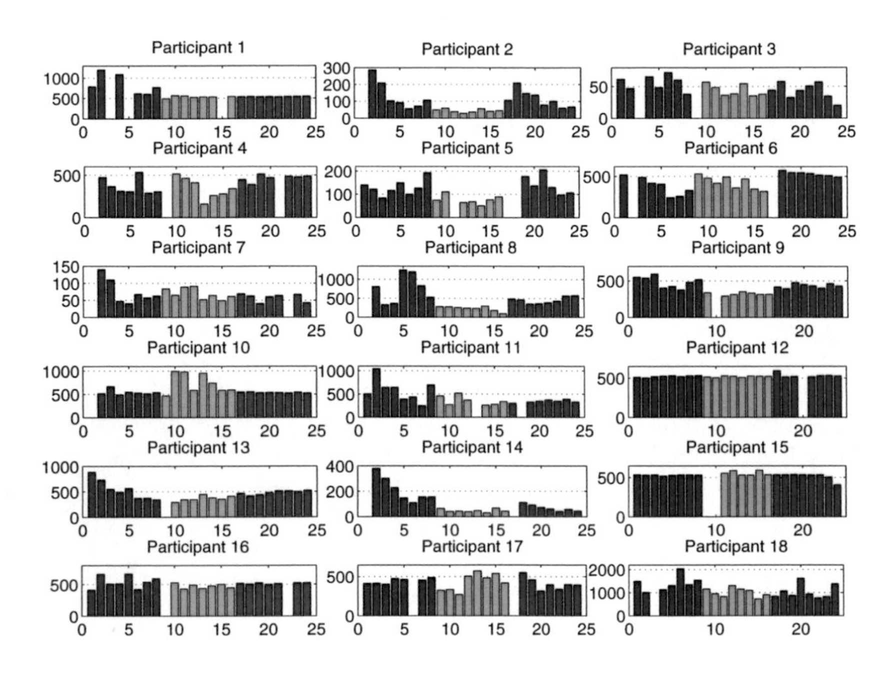

Figure D.1. CAAF Experiment detailed results. Find in the vertical axes the IAE about the task altitude. The missing bars refer to trials in which the aerodynamic stall occurred (non-linear aircraft dynamics and naïve participants, i.e. not professional pilots, were employed in this experiment).

according to its Phugoid mode, are simulated by artificially injecting three control disturbances (elevator impulses) of randomized duration (2, 3 or 3.5 seconds), starting time and sign (upwards or downwards).

The experiment consisted of three different force conditions: No Force condition (referred as *NoEF* condition) with only the spring-damper force on the end-effector, *IHA* condition (the Force Injection CAAF from Equation (3.26) and Equation (3.21)) and *DHA* condition (force from Equation (3.29) and Equation (3.28)).

All the trials (36 of 60 seconds each, 12 trials per condition) were mixed and counter-balanced according to the Latin Squared Method to test natural reaction of the pilots to the three different conditions. Before starting the experiment, every pilot was asked to run a 5 minutes trial where he/she had to perform a slightly different altitude regulation task; the goal of this initial trial, was to let the pilot acquire enough knowledge of aircraft dynamics to be able to confidently pilot it. During this trial a simple spring-damper (stiffness and damping constants were chosen as 1/6 of the NoEF case) behavior of the stick was employed. In total the experiment lasted 90 minutes (including instructions and breaks).

Pilot No.	**1**	**2**	**3**	**4**	**5**	**6**	**7**
Trial No.1	1	3	2	2	1	3	3
Trial No.2	3	1	3	2	3	3	1
Trial No.3	2	2	3	2	1	1	2
Trial No.4	1	3	2	3	2	2	2

Table D.2. The blocks order of presentation for each of the 7 professional pilots. 1: *NoEF*; 2: *IHA*; 3: *DHA*.

No instructions were given about the force conditions to test natural reaction of the pilots to the three different conditions.

D.2 shows an example about the force conditions scheduled for 4 over 36 trials and for the 7 pilots.

In order to focus on the haptic cue the experiment was made more difficult for the professional pilots by setting the Artificial Horizon inoperable (zero pitch and roll).

In each trial there were 3 impulses of 3 different randomized (Latin Square Method) amplitudes (2, 3, 4 seconds), at randomized starting times and always the same amplitude (4 cm displacement of the stick) which sign was randomly changed (+/- that is respectively upward or downward wind gust) all counterbalanced in a way that during the 36 trials every subject received the same number of positive and negative disturbances.

As a rule, the first impulse starting time was randomized between 6 and 11 seconds, the second one between 20 and 28 seconds, the third one between 40 and 46 seconds. As long as the time between each impulse and the next one was randomized between 14 and 23 seconds, after every impulse there was enough time to re-establish the trim conditions.

By using for each trial counter-balanced force conditions as in Table D.2 and similar planned amplitude, starting times and sign impulses to simulate the wind gusts, no learning about the impulses was ensured.

D.2.1 Instructions to professional pilots

You are going to pilot a simulated aircraft through the use of the Omega Device which is a force feedback device, i.e. when you move its end-effector you can feel a force feedback. During the experiment you will watch at the electronic instrument display: on the right side you see the altitude, in the left side the airspeed, in the center the artificial horizon set as inoperable. The only dynamics you have to control is the longitudinal dynamics climbing or diving only. The only needed commands are forwards or backwards (as in a typical aircraft control bar). In each trial there will be 3 vertical wind gusts of random duration, at randomized starting times and of randomized sign (upwards or downwards). The task of the experiment is to fly leveled in trimmed condition at constant altitude ($300ft$) although the presence of the randomized wind gusts by watching the altimeter only. In fact, as said, the Artificial Horizon is set inoperable. Before starting the real experiment you will run a 5 minutes trial to familiarize with the setup. During this trial, you have to fly at the altitude suggested by the magenta marker: at the starting point you have to fly horizontally (0 degrees in the artificial horizon and 300 ft altitude), after about 10 seconds you have to fly at 310 ft altitude, after about 40 seconds you have to fly at 290 ft altitude, after about 40 seconds you have to fly at 300 ft altitude as in the initial conditions and so on till 5 minutes. You have just to follow the magenta marker which will move from one desired altitude to the following one. You are going to run 36 trials of 60 seconds each. In total the experiment will last 90 minutes. At the end of the whole experiment you have some question to answer.

D.2.2 Subjects detailed results

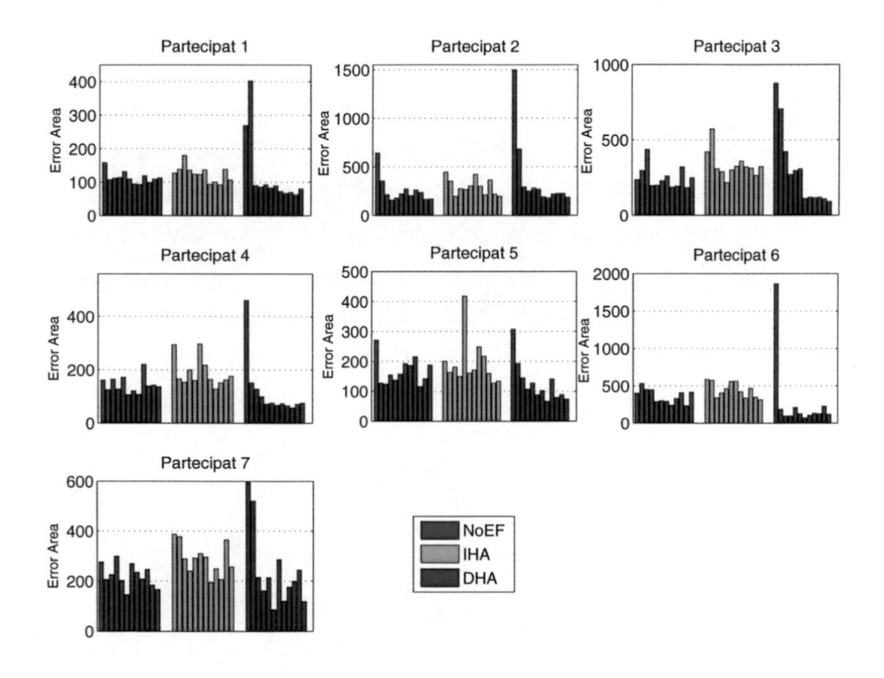

Figure D.2. The CAAF VS DHA Experiment detailed results. Find in the vertical axes the IAE about the task altitude.

In Figure D.2, in each horizontal axis the 3 types of force were grouped according to the legend colors.

The correspondence with the results provided in Section 3.6.4 is evident.

D.3 The OAF VS DHA Experiment

In order to test the IHA-Obstacle Avoidance concept (described in Chapter 4), several experiments about an obstacle avoidance task were run.

Ten naïve subjects participated in the experiment.

A simple control task was prepared: the aircraft had to be flown in an urban canyon with buildings placed irregularly (non Manhattan-like) along the desired path; thus, the buildings constituted a narrow street with buildings in both sides. The task of the experiment was to get the end of the street by avoiding the collisions with them. Five different scenarios (i.e. position of the N obstacles) were used to avoid the effect of learning in test subjects. An example about one of the five employed scenarios is depicted in Figure D.3.

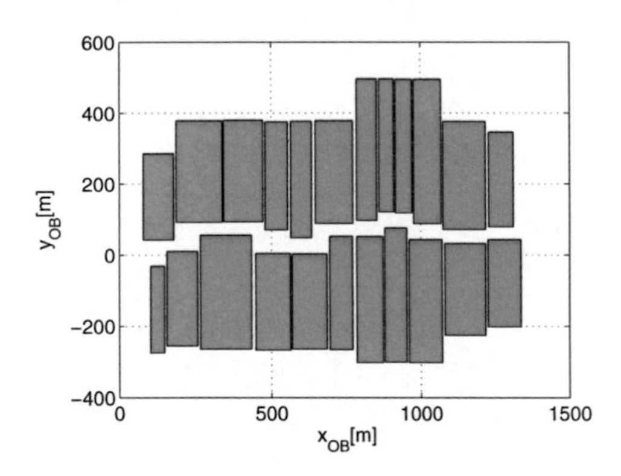

Figure D.3. Aerial projection about one of the five employed scenarios.

To test the natural response to the different types of force no instructions were given to the participants about the force they were going to feel on the stick.

The experiment consisted of three different force conditions and three different fog conditions. The force conditions are: No Force condition (referred as *NoEF*) with only the spring-damper force on the end-effector, *IHA* condition (the IHA-OAF from Section 4.4.3) and *DHA* condition (force from Section 4.4.2). The fog conditions are pretty good (*A*), medium (*B*) and very poor (*C*) visibility conditions.

All the trials (45 of 120 seconds each, 5 trials per force and visibility condition) were mixed and counter-balanced according to the Latin Squared Method to test natural reaction of the pilots to the three different conditions. In total the experiment lasted about 120 minutes.

D.3.1 Instructions to subjects

> You are going to pilot a simulated aircraft through the use of the Omega Device which is a force feedback device, i.e. when you move its end-effector you can feel a force feedback. During the experiment you will watch at the screen in which you will see the scenario of the experiment: a sort of street with buildings in both sides. When the trial starts, you are already flying in the middle of the street and have just to avoid the obstacles on the sides by making turns. The only dynamics you have to control is the lateral one (making the aircraft to turn left or right). To do this you need to move the stick left or right only: forward or vertical end-effector movements do not affect the aircraft trajectory. The task of the experiment is to fly to the end of the street between the buildings by avoiding collisions with them. You will run 45 trials of about 2 minutes each in 3 different fog conditions: the first 15 trials are with pretty good visibility, the second 15 ones are with medium visibility, the third 15 ones are with very poor visibility. During the experiment you will feel through the Omega Device 3 types of forces. One type is only a spring and no aiding force is related to the obstacles. It is similar to the force usually felt on a normal joystick for video-games (when you leave the stick it goes back to the central position). The others two forces are with a sort of force feedback related to the obstacles. We will call them A Force and B Force. These forces instead try to move the stick themselves according to some kind of influence by the obstacles. In all the trials the force conditions are all mixed and after every trial you have to write down which type of force you felt according to your opinion: Spring, A or B Force. You will learn step by step about the A and B Forces and you will be more and more capable of distinguish them identifying some of the differences. When you are ready, you can start each trial by pushing a button on the keyboard. At the end of the experiment you have some question to answer.

In order to avoid the learning factor as much as possible, the types of

	Force Condition	Scenario Type
Trial No.1	1	1
Trial No.2	3	5
Trial No.3	2	4
Trial No.4	3	3
Trial No.5	1	5

Table D.3. Example of planned force conditions and scenario types (from 1 to 5) for each of the 3 visibility conditions and for each of the 10 participants. 1: *NoEF*; 2: *IHA*; 3: *DHA*.

forces, scenarios employed during the experiment and visibility conditions were scheduled according to the Latin Squared Method. An example about the force and scenario plan for first five trials of each of the 10 participants is given in Table D.3.

D.3.2 Subjects detailed results

In Figure D.4, in each horizontal axis the 3 types of force were grouped: blue for NoEF condition, green for IHA condition, red for DHA Force condition.

The correspondence with the results provided in Section 4.5 is evident.

D.4 The MIXED-CAAF/OAF VS DHA Experiment

In order to test the IHA-Mixed CAAF/OAF (described in Chapter 5), several experiments about an obstacle avoidance task in windy conditions were run.

Seven naïve subjects participated in the experiment.

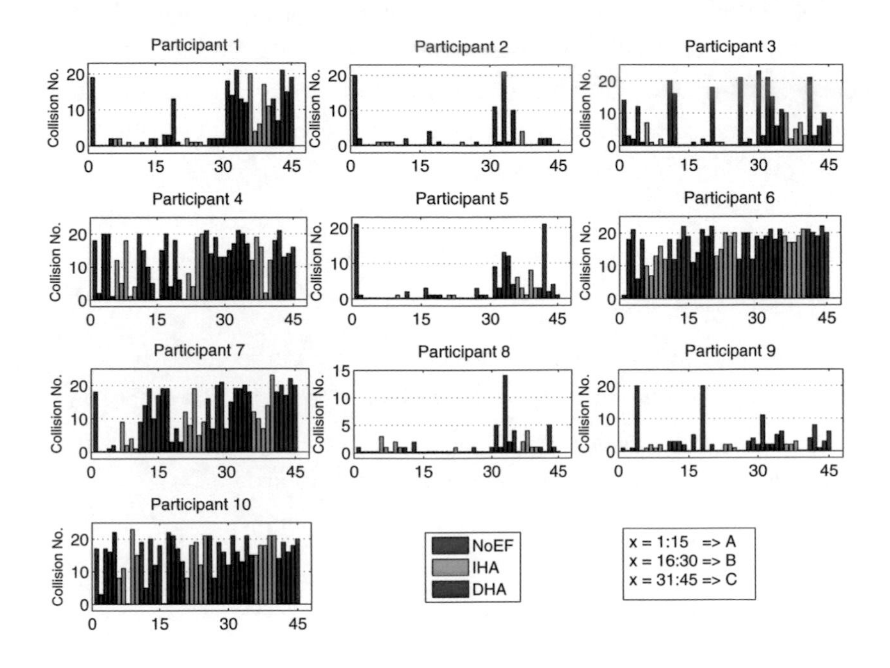

Figure D.4. The Obstacle Avoidance Experiment detailed results (A=Minimum Fog condition; B=Medium Fog condition; C=Maximum Fog condition).

The control task was the same as in the Obstacle Avoidance Experiment: the aircraft had to be flown in an urban canyon with buildings placed irregularly (non Manhattan-like) along the desired path; thus, the buildings constituted a narrow street with buildings in both sides. The task of the experiment was to get the end of the street by avoiding the collisions with them although the presence of 8 lateral wind gusts (4 towards left, 4 towards right). Again five different scenarios (i.e. position of the N obstacles) were used to avoid the learning effect in test subjects.

To test the natural response to the different types of force no instructions were given to the participants about the force they were going to feel on the stick.

The experiment consisted of three different force conditions, two different fog conditions and two different wind conditions. The force conditions are: No Force condition (referred as *NoEF*) with only the spring-damper force on the end-effector, *IHA* condition (the IHA-Mixed CAAF/OAF from Section 5.5.3) and *DHA* condition (force from Section 5.5.2). The fog conditions are medium (*A*) and very poor (*B*) visibility conditions. All the conditions above were run in presence (*W* condition) or absence (*NW* condition) of wind gusts.

All the trials (60 of 120 seconds each, 5 trials per force, visibility and wind condition) were mixed and counter-balanced according to the Latin Squared Method to test natural reaction of the pilots to the three different conditions. In total the experiment lasted about 150 minutes.

D.4.1 Instructions to subjects

You are going to pilot a simulated aircraft through the use of the Omega Device which is a force feedback device, i.e. when you move its end-effector you can feel a force feedback. During the experiment you will watch at the scenario display which depicts a sort of street with buildings in both sides. When the experiment starts, you are already flying in the middle of the street and have just to avoid the obstacles on the sides by making turns. The only dynamics that you have to control is the lateral one (making the aircraft to turn left or right).

To do this you need to move the stick left or right only: forward or vertical end-effector movements do not affect the aircraft trajectory. The task of the experiment is to fly to the end of the street between the buildings by avoiding collisions with them although sometimes, while you will be flying, some sudden lateral wind gust will affect the aircraft. You will run 60 trials of about 2 minutes each. The first 30 trials will be without/with lateral wind gusts (note for the reader: depending on the schedule planned for each subject). The second 30 trials will be with/without lateral wind gusts. There will be two different visibility conditions: a medium visibility condition (some fog is present) and a very poor visibility condition (more fog is present). During the experiment you will feel through the Omega Device 3 types of forces. One type is only a spring type force and no aiding force is related either to the obstacles or to the wind gusts. The other two forces are with a kind of force feedback related to the obstacles and to the wind gusts. We will call them A Force and B Force. In all of the trials the force conditions are all mixed and after every trial you will have to write down which type of force you felt according to your opinion: Spring, A or B Forces. You will learn step by step about the A and B Forces and you will be more and more capable of distinguish them identifying some of the differences. When you are ready, you can start each trial by pushing a button on the keyboard. At the end of the experiment you have some question to answer.

	Force Condition	**Scenario Type**
Trial No.1	3	1
Trial No.2	2	5
Trial No.3	1	4
Trial No.4	3	3
Trial No.5	1	5

Table D.4. Example of planned force conditions and scenario types for one of the 7 participants. 1: *NoEF*; 2: *IHA*; 3: *DHA*.

In order to avoid the learning factor as much as possible, the types of forces, scenarios employed during the experiment, visibility and wind conditions were scheduled according to the Latin Squared Method. An example about the force and scenario plan for the first five trials of one of the 7 participants is given in Table D.4.

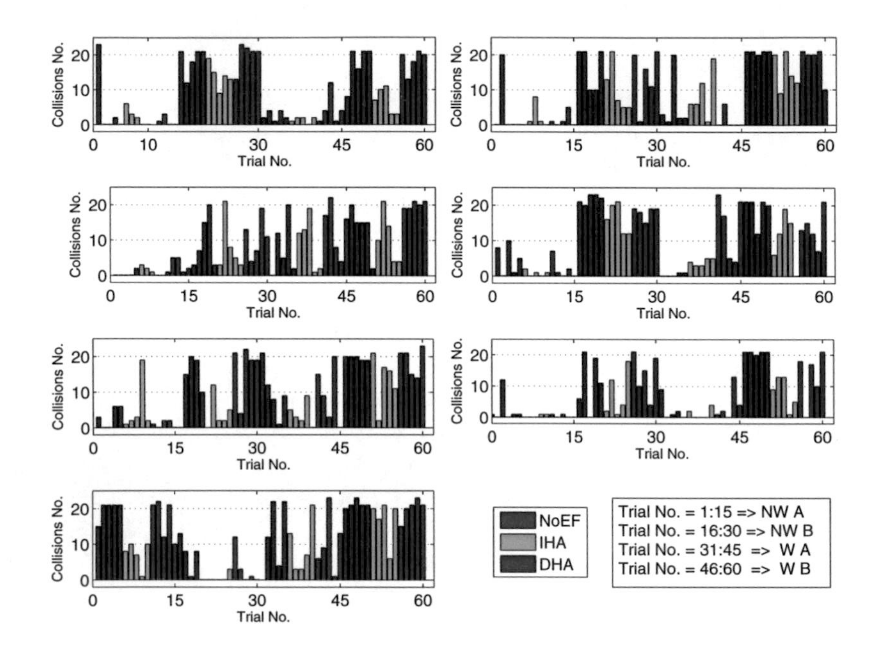

Figure D.5. The MIXED-CAAF/OAF VS DHA Experiment detailed results (NW=No Wind condition; W=Wind condition; A=Minimum Fog condition; B=Maximum Fog condition).

D.4.2 Subjects detailed results

In Figure D.5, in each horizontal axis the 3 types of force were grouped: blue for NoEF condition, green for IHA condition, red for DHA Force condition.

The correspondence with the results provided in Section 5.6 is evident.

Bibliography

[1] Alaimo, S.M.C., Pollini, L., Bresciani, J. P., Bülthoff, H.H., Experimental Comparison of Direct and Indirect Haptic Aids in Support of Obstacle Avoidance for Remotely Piloted Vehicles. *Journal of Mechanics Engineering and Automation*, Volume 2, Number 10, 2012, pp. 628-637.

[2] Alaimo, S.M.C., Pollini, L., Magazzú, A., Bresciani, J.P., Robuffo Giordano, P., Innocenti, M., Bülthoff, H.H. Preliminary Evaluation of a Haptic Aiding Concept for Remotely Piloted Vehicles. In Kappers, A., van Erp, J., Bergmann Tiest, W., van der Helm, F., editors, *Haptics: Generating and Perceiving Tangible Sensations. Proceedings of International Conference, EuroHaptics 2010, Part II, Amsterdam, July 8-10, 2010*, Lecture Notes in Computer Science, pp. 418-425. Springer Berlin Heidelberg New York, 2010. ISBN 3-642-14074-2. Url: `http://dx.doi.org/10.1007/978-3-642-14075-4_62`, DOI: 10.1007/978-3-642-14075-4_62.

[3] Alaimo, S.M.C., Pollini, L., Bresciani, J.P., Bülthoff, H.H., A Comparison of Direct and Indirect Haptic Aiding for Remotely Piloted Vehicles. In Avizzano, C.A., Ruffaldi, E., Carrozzino, M., Fontana, M., Bergamasco, M., editors, *Proceedings of the 19th IEEE International Symposium in Robot and Human Interactive Communication (IEEE Ro-Man 2010), Viareggio (Italy), September 12-15, 2010*, pp. 541-547. IEEE Catalog Number CFP10RHC-CDR. ISBN 978-1-4244-7989-4. Url: http://ieeexplore.ieee.org/xpl/articleDetails.jsp?arnumber=5598647

[4] Alaimo, S.M.C., Pollini, L., Bresciani, J. P., Bülthoff, H.H., Augmented Human-Machine Interface: Providing a Novel Haptic Cueing to the Tele-Operator. *The 3rd Workshop for Young Researchers on Human-Friendly Robotics (HFR 2010), Max Planck Institute for Biological Cybernetics, Tübingen, Germany, October 28th-29th, 2010*. Url: http://hfr2010.wordpress.com/2010/07/06/program/.

[5] Alaimo, S.M.C., Pollini, L., Bresciani, J. P., Bülthoff, H.H., Evaluation of Direct and Indirect Haptic Aiding in an Obstacle Avoidance Task for Tele-Operated Systems. In Bittanti, S., Cenedese, A., Zampieri, S., editors, *Proceedings of the 18th World Congress of the International Federation of Automatic Control (IFAC WC), 28th August - 2nd September, 2011*. Url: http://www.ifac-papersonline.net/Detailed/49613.html

[6] Alaimo, S.M.C., Pollini, J. P., Bülthoff, H.H., Experiments of Direct and Indirect Haptic Aiding for Remotely Piloted Vehicles with a Mixed Wind Gust Rejection/Obstacle Avoidance Task. *Proceedings of the AIAA Modeling and Simulation Technologies Conference 2011, American Institute of Aeronautics and Astronautics, Portland, OR, USA, 8th-11th August, 2011*. Url: http://arc.aiaa.org/doi/abs/10.2514/6.2011-6242?prevSearch=[Fulltext

%3A+[author%3A+Alaimo%2C+Samantha]]&searchHistoryKey=

[7] Alaimo, S.M.C., Pollini, L., Bülthoff, H.H., Admittance-Based Bilateral Teleoperation with Time Delay for an Unmanned Aerial Vehicle involved in an Obstacle Avoidance Task. *Proceedings of the AIAA Modeling and Simulation Technologies Conference 2011, American Institute of Aeronautics and Astronautics, Portland, OR, USA, 8th-11th August, 2011.* Url: http://arc.aiaa.org/doi/abs/10.2514/6.2011-6243?prevSearch=[Fulltext%3A+[author%3A+Alaimo%2C+Samantha]]&searchHistoryKey=

[8] Anderson, R.J., Spong, M.W., Bilateral control of teleoperators with time delay. *Proceedings of the 27th Conference on Decision and Control Austin, Texas, December 1988.*

[9] Artigas, J., Preusche, C., Hirzinger, G., Time domain passivity for delayed haptic telepresence with energy reference. *Proceedings of the 2007 IEEE/RSJ International Conference on Intelligent Robots and Systems, San Diego, CA, USA, Oct 29 - Nov 2, 2007.*

[10] Bicchi, A., Buss, M., Ernst, M.O., Peer, A., The Sense of Touch and its Rendering Progress in Haptics Research, *Springer Tracts in Advanced Robotics*, Volume 45, 2008, DOI: 10.1007/978-3-540-79035-8.

[11] Chopra, N., Spong, M.W., Synchronization of networked passive systems with applications to bilateral teleoperation. In *Society of instrumentation and control engineering of Japan annual conference, Okayama, Japan, August 8-10, 2005.*

[12] Cooke, N.J., Human Factors of Remotely Operated Vehicles. *Proceedings of the Human Factors and Ergonomics Society 50th Annual Meeting*, 2006, pp. 166-169.

[13] Cox, T.H., Nagy, C.J., Skoog, M.A., Somers, I.A., Civil UAV capability

assessment, NASA, December 2004.

[14] David, R.O., Arthur, L.M., Defense Science Board Study on Unmanned Aerial Vehicles and Uninhabited Combat Aerial Vehicles (Office of the Under Secretary of Defense For Acquisition, Technology, and Logistics Washington, D.C. 20301-3140, February 2004), available from `www.acq.osd.mil/dsb/reports/ADA423585.pdf`.

[15] Diolaiti, N., Melchiorri, C., Tele-Operation of a Mobile Robot through Haptic Feedback. *IEEE Int. Workshop on Haptic Virtual Environments and Their Applications (HAVE 2002). Ottawa, Ontario, Canada, 17-18 November 2002.*

[16] Draper, J.V., Kaber, D.B., Usher, J.M., Telepresence, *Human Factors*, Vol. 40, 1998.

[17] Emerson, T.J., Reising, J.M., Britten-Austin, H.G., Workload and situation awareness in future aircraft. SAE Technical Paper (No. 871803). Warrendale, PA: Society of Automotive Engineers, 1987.

[18] Endsley, M.R., Measurement of situation awareness in dynamic systems, *Human Factors*, 1995, 37(1), pp. 65-84.

[19] Endsley, M.R., Farley, T.C., Jones, W.M., Midkiff, A.H., Hansman, R.J., Situational awareness information requirements for commercial airline pilots, *International Center for Air Transportation Department of Aeronautics & Astronautics, Massachusetts Institute of Technology, Cambridge, MA 02139 USA, September 1998.*

[20] Farkhatdinov, I., Ryu, J-H., An, J., A preliminary experimental study on haptic teleoperation of mobile robot with variable force feedback gain, *IEEE Haptics Symposium 2010, 25-26 March, Waltham, Massachusetts, USA.*

[21] Ferrell, W.R., Sheridan, T.B., Supervisory control of remote manipulation, *IEEE Spectrum*, 1967, pp. 81-88.

[22] Franken, M., Stramigioli, S., Reilink, R., Secchi, C., Macchelli, A., Bridging the gap between passivity and transparency, *Proceedings of Robotics: Science & Systems, 2009*.

[23] Furuta, K., Kosuge, K., Shiote, Y., Hatano, H., Master-slave manipulator based on virtual internal model following control concept, *Proceedings of IEEE International Conference on Robotics and Automation, 1987*, pp. 567-572.

[24] Ganjefar, S., Momeni, H., Janabi-Sharifi, F., Teleoperation systems design using augmented Wave-Variables and Smith predictor method for reducing time-delay effects, In *Proceedings of the IEEE international symposium on intelligent control, Vancouver, Canada, 2002*, pp. 333-338.

[25] Gibson, J.C., Hess, R.A., Stick and Feel System Design. Advisory Group for Aerospace Research & Development. AGARDograph 332. Canada Communication Group, Hull, Canada (1997).

[26] Goertz, R., Electronically controlled manipulator, *Nucleonics*, Vol. 12, pp. 46-47, November 1954.

[27] Hannaford, B., Ryu, J.H., Time-Domain Passivity Control of Haptic Interfaces, In *IEEE Transaction on Robotics and Automation*, Vol. 18, No. 1, 2002.

[28] Hess, R.A., Theory for Aircraft Handling Qualities Based Upon a Structural Pilot Model, *Journal of Guidance, Control, and Dynamics*, Vol. 12, No. 6, 1988, p. 792.

[29] Hing, J.T., Oh, P.Y., Development of an Unmanned Aerial Vehicle Piloting System with Integrated Motion Cueing for Training and Pilot

Evaluation, *Journal of Intelligence Robot Systems*, 2009, Vol. 54, pp. 3-19. DOI: 10.1007/s10846-008-9252-3.

[30] Hirche, S., Buss, M., Transparent data reduction in networked telepresence and teleaction systems, Part II: Time-delayed communication. *Presence: Teleoperators & Virtual Environments*, 2007, 16(5), pp. 532-542.

[31] Hokayem, P.F., Spong, M.W., Bilateral teleoperation: an historical survey, *Automatica*, 2006, Vol. 42, pp. 2035-2057.

[32] Horan, B., Creighton, D., Nahavandi, S., Jamshidi, M., Bilateral haptic teleoperation of an articulated track mobile robot, *Proceedings of IEEE International Conference on System of Systems Engineering, SoSE '07, San Antonio, TX, 2007.*

[33] Horan, B., Najdovski, Z., Nahavandi, S., Multi-point Multi-hand Haptic Teleoperation of a Mobile Robot, *The 18th IEEE International Symposium on Robot and Human Interactive Communication, Toyama, Japan, Sept. 27-Oct. 2, 2009.*

[34] Hosman, R.J.A.W., Benard, B., Fourquet, H., Active and passive side stick controllers in manual aircraft control. *Proceedings of IEEE International Conference on Systems, Man and Cybernetics, 4-7 Nov 1990*, pp.527-529. DOI: 10.1109/ICSMC.1990.142165.

[35] James, T., Multi-mission/multi-agency reconfigurable UAV, *Unmanned Systems*, Winter, 1994, pp. 41-42.

[36] Jones, D.G., Endsley, M.R., Sources of situation awareness errors in aviation. *Aviation, Space and Environmental Medicine*, 1996, 67(6), pp. 507-512.

[37] Kim, W.S., Experiments with a predictive display and shared compliant control for time-delayed teleoperation. In *Proceedings of*

the annual international conference of the IEEE engineering in medicine and biology society, 1990, pp. 1905-1906.

[38] Kveraga, K., Boucher, L., Hughes, H.C., Saccades operate in violation of Hick's law, *Exp Brain Res.*, October 2002, 146(3), pp. 307-314. Published online 2002 August 10. DOI: 10.1007/s00221-002-1168-8.

[39] Lam, T.M., Artificial Force Field for Haptic Feedback in UAV Teleoperation, Ph.D. thesis, Faculty of Aerospace Engineering, Delft University of Technology (TU Delft), Delft, The Netherlands, 2009.

[40] Lam, T.M., Boschloo, H.W., Mulder, M., van Paassen, M.M., Artificial Force Field for Haptic Feedback in UAV Teleoperation. In: *IEEE Transactions on Systems, Man and Cybernetics, Part A: Systems and Humans*, Nov. 2009, Vol. 39, Issue 6, pp. 1316-1330.

[41] Lam, T.M., Mulder, M., van Paassen, M.M., Collision Avoidance in UAV Tele-operation with Time Delay. *Systems, Man and Cybernetics*, 2007. *IEEE International Conference on ISIC*, pp. 997-1002, 7-10 Oct. 2007. DOI: 10.1109/ICSMC.2007.4413867.

[42] Lam, T.M., Mulder, M., van Paassen, M.M., Haptic Interface For UAV Collision Avoidance, *The International Journal of Aviation Psychology*, 17(2), pp. 167-195.

[43] Lam, T.M., Mulder, M., van Paassen, M.M., Mulder, J.A., van Der Helm, F.C.T., Force-stiffness Feedback in UAV Tele-operation with Time Delay. In *AIAA Guidance, Navigation, and Control Conference, Chicago, Illinois, August 2009.*

[44] Lawrence, D.A., Stability and transparency in bilateral teleoperation, In *IEEE Transactions on Robotics,and Automation*, 1993, Vol. 9, pp. 624-637.

[45] Lippay, A.L., Kruk R., King, M., Murray, M., Flight Test of a Displacement Sidearm Controller. *Annual Conference of Manual Control*, 17 June 1985.

[46] Ljung, L., *System Identification: Theory for the User*, second edition, Prentice Hall, New Jersey, 1999.

[47] McCarley, J.S., Wickens, C.D., Human Factors Implications of UAVs in the National Airspace, Institute of Aviation Aviation Human Factors Division University of Illinois at Urbana-Champaign, available from `http://www.tc.faa.gov/logistics/grants/pdf/2004/04-G-032.pdf`.

[48] McRuer, D., Weir, D.H., Theory of Manual Vehicular Control, *IEEE Transactions on Man-Machine Systems*, Dec. 1969, Vol.10, No.4, pp.257-291. DOI: 10.1109/TMMS.1969.299930.

[49] Miyazaki, F., Matsubayashi, S., Yoshimi, T., Arimoto, S., A new control methodology towards advanced teleoperation of master-slave robot systems, *Proceedings of IEEE International Confonference on Robotics and Automation*, 1986, Vol.3, pp. 997-1002.

[50] Montano, F., Integrazione di Active Sticks nell'architettura fly-by-wire dell'Alenia Aermacchi M-346, Universita' degli Studi di Palermo, Palermo, Italy, 2006.

[51] Mouloua, M., Gilson, R., Kring, J., Hancock, P., Workload, situational awareness, and teaming issues for UAV/UCAV operations, *Proceedings of the Human Factors and Ergonomics Society 45th Annual Meeting, 2001*, pp. 162-165.

[52] Niemeyer, G., Slotine, J-J.E., Telemanipulation with time delay, *The International Journal of Robotics Research*, 2004, Vol. 23, pp. 873. DOI: 10.1177/0278364904045563.

[53] Nisser, T., Westin, C., Human Factors challenges in Unmanned Aerial Vehicles (UAVs): a literature review, Lund University School of Aviation, Ljungbyhed, Sweden, 2008.

[54] Nordh, R., Berrezag, A., Dimitrov, S., Turchet, L., Hayward, V., Serafin, S., Preliminary experiment combining virtual reality haptic shoes and audio synthesis, *Proceedings of International Conference, EuroHaptics 2010, Part I, Amsterdam, July 2010*, pp. 123-129.

[55] Oliver, D. R. and Money, A. L. (2001). Unmanned Aerial Vehicles Roadmap. Technical Report, Department of Defense, Washington DC.

[56] Peer, A., Design and Control of Admittance-Type Telemanipulation Systems, Ph.D. thesis, Institute of Automatic Control Engineering, Technische Universität München, 2008.

[57] Peer, A., Buss, M., A New Admittance Type Haptic Interface for Bimanual Manipulations. *IEEE/ASME Transactions on Mechatronics*, 2008, 13(4), pp. 416-428.

[58] Pollini, L. Innocenti, M., A synthetic environment for dynamic systems control and distributed simulation, *IEEE Control Systems Magazine*, April 2000, Vol 20, Num. 2, pp. 49-61.

[59] Ren, W., Beard, R.W., Satisficing approach to human-in-the-Loop safeguarded control, American Control , Portland, OR, USA, June 8-10, 2005.

[60] Robuffo Giordano, P., Deusch, H., Lächele, J., Bülthoff, H.H., Visual-Vestibular Feedback for Enhanced Situational Awareness in Teleoperation of UAVs, *Proceedings of the American Helicopter Society 66th Annual Forum and Technology Display*, 1-10 May 2010, AHS International, Alexandria, VA, USA.

[61] Royal Aeronautical Society (RAES), Human Factors Group, available from http://www.raes-hfg.com/crm/reports/sa-defns.pdf.

[62] Tadema, J., Theunissena, E., Koenersb, J., Using perspective guidance overlay to improve UAV manual control performance, *Enhanced and Synthetic Vision*, 2007, edited by Jacques G. Verly, Jeff J. Guell, *Proceedings of SPIE*, Vol. 6559, 65590C.

[63] Tanner, N.A., Niemeyer, G., Online tuning of wave impedance in telerobotics, *Proceedings of the 2004 IEEE Conference on Robotics, Automation and Mechatronics*, Singapore, 1-3 December, 2004.

[64] Tenney, Y.J., Adams, M.J., Pew, R.W., Huggins, A.W., and Rogers, W.H., A principle approach to the measurement of situation awareness in commercial aviation. NASA contractor report 4451, Langley Research Center: NASA, 1992.

[65] de Vlugt, E., Identification of Spinal Reflexes. Ph.D. dissertation, Faculty of Design, Engineering, and Production, Delft University of Technology, Delft, The Nederlands, 2004.

[66] de Vries, S.C, UAVs and control delays, TNO report, TNO Defence, Security and Safety, September 2005.

[67] Raju, G. J., Verghese, G. C., Sheridan, T. B., Design issues in 2-port network models of bilateral remote manipulation. In *Proceedings of the IEEE international conference on robotics and automation, 1989*, Vol. 3, pp. 1316-1321.

[68] Rauw, M.O., FDC 1.2 - A Simulink Toolbox for Flight Dynamics and Control Analysis, Zeist, The Netherlands, 1997 (second edition: Haarlem, The Netherlands, 2001). Distributed exclusively across the Internet http://www.dutchroll.com.

[69] Roskam, J., *Airplane Flight Dynamics and Automatic Flight Controls*

Part I, DARcorporation Design, Analysis, Research, 120 East 9th Street, Suite 2, Lawrence, Kansas 66044, U.S.A, 2001.

[70] Ruff, H.A., Draper, M.H., Lu, L.G., Poole, M.R., Repperger, D.W., Haptic feedback as a supplemental method of alerting UAV operators to the onset turbulence, *Proceedings of the IEA 2000/ HFES 2000 Congress*, 3.41 - 3.44.

[71] Sangyoon L., Sukhatme, G.S., Kim, G.J., Chan-Mo P., Haptic control of a mobile robot: a user study, *Proceedings of IEEE/RSJ International Conference on Intelligent Robots and Systems*, 2002, Vol.3, pp. 2867-2874. DOI: 10.1109/IRDS.2002.1041705.

[72] Schauss, T., Vittorias, I., Passenberg, C., Peer, A., Buss, M., Tutorial for Telerobotic Summer School 2010 - Control Group, Institute of Automatic Control Engineering, Technische Universität München, July 26-30, 2010, Munich, Germany.

[73] Schmidt, A., Lee, D., *Motor Control and Learning, A behavioral Emphasis*, 4th Ed., Human Kynetics, 2005.

[74] Sheridan, T.B., Space teleoperation through time delay: review and prognosis, *IEEE Transactions on Robotics and Automation*, Oct 1993, Vol.9, No.5, pp.592-606. DOI: 10.1109/70.258052.

[75] Sheridan, T.B., Ferrell, W.R., Remote Manipulative Control with Transmission Delay, *IEEE Transactions on Human Factors in Electronics*, Sept. 1963, Vol. HFE-4, Issue 1, pp. 25-29.

[76] Sheridan, T.B., Teleoperation, telerobotics and telepresence: a progress report. *Control Engineering Practice*, 1995, 3(2), pp. 205-214. *Human Factors*, 1998, Vol. 40.

[77] Sheridan, T.B., *Telerobotics, automation, and human supervisory control*, Cambridge, MA, The MIT Press, 1992.

[78] Stevens, B.L., Lewis F.L., *Aircraft Control and Simulation*, 2nd ed., John Wiley & Sons Inc., 111 River Street, Hoboken, New Jersey 07030, U.S.A, 2003.

[79] de Stigter, S., Mulder, M., van Paassen, M.M., Design and Evaluation of a Haptic Flight Director, *Journal of Guidance, Control, and Dynamics*, January-February 2007, Vol. 30, No. 1.

[80] Stramigioli, S., Secchi, C., van der Schaft, A.J., Fantuzzi, C., Sampled data systems passivity and discrete Port-Hamiltonian systems, *IEEE Transactions on Robotics*, August 2005, Vol. 21, No. 4.

[81] Mark B. Tischler, Advances in Aircraft Flight Control, Ed. (London, UK: Taylor & Francis, 1996).

[82] Mayer, J., Cox, T.H., Evaluation of Two Unique Side Stick Controllers in a Fixed-Base Flight Simulator, NASA Dryden Flight Research Center Edwards, California.

[83] Reichenbach, A., Thielscher, A., Peer, A., Bülthoff H.H., Bresciani, J-P. (2009), Seeing the hand while reaching speeds up on-line responses to a sudden change in target position. *The Journal of Physiology*, 587(19), pp. 4605-4616.

[84] Van Erp, J.B.F., Van Veen, H.A.H.C., Jansen, C., Dobbins, T., Waypoint Navigation with a Vibrotactile Waist Belt, *ACM Trans. Appl. Percept.*, 2(2), pp. 106-117.

[85] Yokokohji, Y., Imaida, T., Yoshikawa, T., Bilateral control with energy balance monitoring under time-varying communication delay. In *Proceedings of the IEEE international conference on robotics and automation, San Francisco, CA, USA*, Vol. 3, pp. 2684-2689.

[86] Wickens, C.D., *Engineering Psychology and Human Performance*, 2nd ed., New York, Harper Collins, 1992.

[87] Zhu, W-H., Salcudean, S.E., Stability guaranteed teleoperation: an adaptive motion/force control approach, *IEEE Transaction On Automatic Control*, 2000, Vol. 45, No. 11, pp. 1951-1969.